Simon,

Warmest regards!

John

CAUSATION IN PSYCHOLOGY

CAUSATION IN PSYCHOLOGY

JOHN CAMPBELL

 Harvard University Press
Cambridge, Massachusetts
London, England
2020

Library of Congress Cataloging-in-Publication Data
Names: Campbell, John, 1956- author.
Title: Causation in psychology / John Campbell.
Description: Cambridge, Massachusetts : Harvard University Press, 2020. |
    Includes bibliographical references and index.
Identifiers: LCCN 2020017718 | ISBN 9780674967861 (cloth)
Subjects: LCSH: Technology and civilization. | Causation. | Robots. |
    Artificial intelligence. | Empathy.
Classification: LCC CB478 .C294 2020 | DDC 303.48/3—dc23
LC record available at https://lccn.loc.gov/2020017718

for Rory, light of my life

CONTENTS

CAUSATION IN PSYCHOLOGY

INTRODUCTION

GENERAL VS. SINGULAR CAUSATION

On the one hand, there are general causal claims, anything of the same form as

"Humiliation causes depression"

or

"Salt causes heart disease,"

where we seem to be dealing with a relation between two variables, such as humiliation or depression, which can take any of a range of values. In contrast, there are singular causal claims, such as

"Billy's desire for revenge caused him to attack"

or

"Sally's high salt intake caused her death from heart disease."

These singular causal claims seem to relate not variables but names of particular, concrete events.

The relation between general and singular causal claims is puzzling in any domain. For example, it's a natural idea that general causation involves some kind of quantification over cases of singular causation. You might suggest that "X causes Y" means something like, "There are many cases in which an instance of X causes an instance of Y." But as Christopher Hitchcock once pointed out, it does not on the face of it seem that general causal claims can be regarded as somehow quantifying over singular causal claims. If you take a remark like

"Eating a pound of Uranium-235 causes death,"

presumably that's true, even if no one will ever eat a pound of Uranium-235. Therefore, it's hard to see what kind of collection of cases of singular causation a claim of general causation might be thought to reflect.

There is actually a deeper reason why singular causation may resist assimilation to general causation. If you think of causation in terms of generalizations, then it seems there can't be any more to singular causation than the instantiation of general causation. But suppose you think of causation in terms of *processes* connecting cause and effect. A paradigm might be the trail of gunpowder connecting the lighting of a match to the detonation of a bomb. There might not be any generalization to be had about the connection between lighting matches and explosions. Sometimes the match lights, and there is an explosion; sometimes the match lights, and there isn't. There might also be nothing general to be said about what kind of fuse is required for the connection to be made, other than the vacuous, "It has to be something that can causally connect the lighting of a match to an explosion." Nonetheless, in any particular case, we can follow the path from the lighting of the match to the detonation of the bomb, and in any one case, we can know decisively what was the cause of the detonation.

Of course, one might argue that the operations of processes themselves are to be understood in terms of generalizations. But it is not obvious how this is to be done in the physical case, and it is still less obvious how this might be done in the psychological case given the welter of factors that may figure in a psychological process and the difficulty of finding generalizations governing how they might interact with one another to generate an outcome.

Consider the relation between "Humiliation causes depression" and "Sally's humiliation caused her depression." The mere fact that you've established the general causal claim does not show that the singular claim is correct. It could be that humiliation causes depression, that Sally had been humiliated and had gotten depressed, but that the humiliation was no part of the cause of the depression. Similarly, in the physical case, it could be that salt does cause heart disease, that Sally did take a lot of salt, that she did get heart disease, but that this was not because of the salt but for some other reason. We do have some understanding of how an autopsy might establish that salt caused Sally's heart disease. But how could we establish that it was Sally's humiliation that caused her depression? Well, how do we do it in ordinary life? If you are Sally's therapist or friend, you may be able to imaginatively follow her chain of thoughts and feelings from the initial episode. You may be able report quite conclusively on how the chain of causation went through to the onset of depression. That is part of what we do in everyday life.

We have a conception of psychological process that is central to our understanding of the dynamics of our mental lives. This is the psychological analog of the trail of gunpowder, and I'll be trying to explain how we have this conception of singular psychological causation and indicate why it is central to our picture of human freedom and what matters in ordinary life.

It should not be taken for granted why we make singular causal claims at all—and given that we do, why we assign them the importance that we do. Humans seem to be unique among animals in doing this. Most animals seem to have no representations of causation at all, and there seems to be none apart from humans that has been demonstrated to think in terms of singular causation.

To see the puzzle, suppose we find a people somewhere in a sequestered region of the planet who are intelligent (and perhaps they even speak English) to the extent possible given the following stipulation: they have no concept of singular causation. Perhaps they do have a concept of general causation, and they have run randomized controlled trials to determine the outcomes of various practices. They may therefore have enforced norms of social behavior. They have a general concept of good practice: it's good practice to take some vitamin E each day, at-risk people should be administered tamoxifem, and so on. This can all be justified within talk about general causation itself, given that they know what kinds of outcomes they regard as desirable and which as not. They are all rigorously trained to wash their hands before eating, and so on. But they simply don't have our concept of singular causation.

Suppose we talk to these people and try to persuade them that they should do things our way. They're intelligent and reasonable. But when we explain to them about singular causation, the ascription of responsibility, and so on, they don't see why they should introduce any such concept, and they certainly don't see why they should organize their practical reactions to one another around any such concept. Is there a quick, compelling explanation why they should shift to doing things our way? Though I won't pursue this here, I suspect that our uncertainty as to the motivation for talking about singular causation in the first place is part of the reason why we have difficulty over topics such as moral luck or the trolley problem, which seem to turn on relatively subtle differences between types of singular causation. To take one example, suppose Suzy and Billy are throwing rocks at a bottle. If Suzy hits the bottle as Billy's rock whistles through empty air a moment later, we blame Suzy but not Billy for the breaking of the bottle. This is a classic case of moral luck. What was in Suzy's heart was, we can suppose, pretty much the same as what was in Billy's heart. Nonetheless, Suzy had the moral bad luck; she is more to blame than Billy because she broke the bottle. But from the viewpoint of our hypothetical people, who are concerned only with general causation and good practice, Suzy and Billy's behaviors seem exactly on a par. Our hypothetical people do not understand why it's right to differentiate between Suzy and Billy in our responses to what they did, and it's not easy to know how we should go about explaining this to them.

It's instructive here to reflect on the understanding of causation had by nonhuman animals. Even the most sophisticated demonstrations of causal reasoning by nonhuman animals so far do not of themselves suggest that animals are capable of establishing singular causes, as opposed to finding general causal patterns. And though it's relatively easy to see how one might test for animal grasp of general causation, it's hard to see how one would establish that an animal grasps singular causation. The trouble is this. Suppose we observe the animal reacting to a case in which one event, A, caused another, B. Has the animal spotted singular causation, or has it merely drawn evidence for a causal connection between one of the variables, X, characterizing the cause event and one of the variables, Y, characterizing the outcome event? The natural place to look is at the subsequent behavior of the animal. It's possible the subsequent behavior of the animal will show grasp of a causal connection between X and Y; the animal may try to affect Y by changing X, for example (cf. Taylor, Miller, and Gray 2012; Blaisdell et al. 2006, 2010). In that case, we'd seem to have behavioral evidence that the animal grasped the general causal connection between X and Y. But what behavioral evidence would we look for to establish that the animal had grasped the singular relation, that event A had caused event B? In human life, there are some natural things we could look for. If I think that it was your foot that caused me to trip, for example, I might exhibit resentment. This kind of thing isn't a matter of my drawing causal lessons for the future. But it's hard—not impossible, but hard—to see how you would try to find this kind of thing in animals. Of course, an animal pelted by a conspecific might be keen to establish which had pelted it; but the natural interpretation of what is going on here is that the animal is trying to find evidence for causal generalizations governing one conspecific rather than another.

The point here is that among animals, it seems entirely possible that we could find a grasp of general causation without a grasp of singular causation. At the moment, that seems to be what we do find. It's therefore hard to see how we could argue that a grasp of general causation depends on a grasp of singular causation. So why does it help that we have an understanding of singular causation? It is at the center of our moral and practical lives, our habits of praise and blame, our resentment or gratitude—all of social life. I will propose that it's distinctive of human psychology that we have a singular psychological causation that's not grounded in general causation, and that this is central to our conception of ourselves as free. And it's in the need to coordinate the social lives of creatures that have this singular causation in their psychological lives, not grounded in causal generalizations, that we find much of what is most distinctive about human life.

How do we know about causation? We establish general causal claims in the mental in the very same way that as we establish general causal claims in the physical. Suppose we want to find what a drug does. We divide our population into two cohorts, one that gets the drug and one that doesn't, and we look for a difference in outcome across the two groups to see what the drug does. Suppose we want to find whether cognitive behavioral therapy works for insomnia. We again divide our population into two groups, one of which we give the treatment to and the other we don't, and we look for the difference in outcome across the two groups to find what the therapy does. Of course there are always methodological problems, but the principles being used seem to be the same whether we are dealing with physical or with mental causation, so it seems quite difficult to argue that the mental is somehow not causal. And in fact, this way of finding causation is ubiquitous in the social sciences, clinical psychology, and psychiatry.

That's true for causation at the general level, at which we think of it as a relation between variables (such as cognitive-behavioral therapy and insomnia). Much of the work that scientists do on establishing causal connections is to establish general causal truths, such as "smoking causes cancer" or "humiliation causes depression." As we saw, though, we can distinguish between causation at this general level and singular causation, which is a relation between particular concrete events (cf. Eells 1991). It's a further step from saying "Smoking causes cancer" to saying "Sally's smoking caused her death from cancer." It could be that Sally smokes and that Sally died from cancer, but that the two things were not causally connected and that the cause of her death from cancer was something else altogether. Similarly, it can be that humiliation causes depression, that Ting-An was humiliated, and that Ting-An got depression, even though her humiliation was not the cause of her depression.

In the case of a randomized controlled trial, we can regard the knowledge of what causes what as being derived from knowledge of probabilities. We do the experimental intervention and look at the numbers as to what happened, to provide the basis for our assessment of causality. In the singular case, however, the thing often seems to go round the other way. Our knowledge of causal connections is the basis for our assessment of probabilities. Suppose we take a classic example of singular causation from I. J. Good (1961–1962): Moriarty and Watson at Reichenbach Falls. Moriarty is on the cliff top and is about to roll a rock over the edge onto Holmes, who is strolling on the waterfront below. Moriarty is fiendishly dexterous and stands a very good chance of getting Holmes. At the last moment, just as the rock is about to go over, Watson rushes up. There isn't time to tackle Moriarty. He gives the rock a wild shove, and it topples haphazardly from side to side down the rugged cliff face, eventually crushing Holmes. Good's idea was that this is a case in which Watson's push actually lowered the probability of Holmes being crushed; nonetheless, it's evident that Watson's push did actually cause Holmes's death. The key epistemic point, though, is that for you as an observer watching this scene, it's absolutely evident, and absolutely compelling, that Watson's push caused Holmes's death. You followed the trajectory of the boulder from the shove to the crush. Did Watson's push make Holmes's death more likely? Insofar as you do have any knowledge of probabilities, it is derived from your knowledge of causation. The epistemic situation in the singular case is therefore quite different to the situation in a randomized controlled trial, where we really are trying to derive knowledge of causation from knowledge of probabilities. The same point holds in the mentalistic case. When you see Hanjo's reaction to Sally coming into the room, you follow his train of thought, and you know what is causing what. But you may be quite unsure about the single-case probabilities here. Insofar as you do have any knowledge of relevant probabilities, it is derived from your knowledge of the singular causal relations.

We therefore need to know something about what singular mental causation is so we can see how our ordinary imaginative understanding of one another provides knowledge of it. What is distinctive about the humanities is the use of the imagination to track the ballistics of people's thoughts and feelings.

Let's locate these points with respect to a classical argument for the existence of mental causation. Davidson (1980) argued against philosophers such as Ryle and Wittgenstein, who held that mentalistic explanation was a matter of "rationalizing" the subject's actions, fitting them into a narrative of the mental life that made sense without invoking any idea of causation. Davidson's argument was quite simple. Someone can have more than one reason to perform an action. Consider Brutus at the assassination of Caesar. Suppose Brutus was jealous of Caesar, to the point of wanting Caesar dead. Suppose Brutus also loves the republic and wants Caesar dead in order to restore it. Brutus has both motives. Both "rationalize" his action perfectly well. So on the Ryle / Wittgenstein picture, we seem to have told the whole explanatory story when we have specified both motives. Davidson's point was that although one can have more than one reason to perform an action, it may still be true that only one reason was operative— that the person performed the action *for* one reason rather than another.

In explaining Brutus's action, we give a lot of weight to the question "Did he assassinate Caesar because he was jealous, or did he assassinate Caesar because he loved the republic?" Whether Brutus is executed as a criminal may turn on the answer we give to that question. But when we try to pinpoint what question we are asking here, it seems compelling that it is a causal question.

The question is, Which of these motives that Brutus certainly had is the one that *caused* Brutus to kill Caesar? This argument for treating reasons as causes was found compelling at the time, and philosophers today still generally regard it as conclusive in favor of a causal account of explanation by reasons. It does happen that people have more than one reason to perform an action and that we give weight to the question "Which is the reason *for which* this person action?" This seems to be the same as the question "Which reason *caused* this person to act?"

What concept of causation do we need here? We can't think of causation as a mere matter of counterfactual dependence ("If he hadn't been jealous, then Caesar wouldn't have died"). For on a natural reading of it, this is a case in which

a. Brutus acted from one motive rather than the other, yet
b. if he hadn't acted from that motive, the other would have spurred him on to do the thing anyway.

The intuitive idea is that there's a process connecting one motive rather than the other to the action. But what concept of process is to the point here? Davidson himself had no answer to this: he did not recognize the importance of the notion of process in an account of causation, either at the mental or at the physical level, but he thought that the analysis of causation could be achieved entirely in terms of exceptionless laws. He also thought that these laws operate at the level of physical characteristics. The first problem with this picture is to relate it to a persuasive account of how we actually find out about these matters. When Brutus's countrymen found him to be acting from one motive rather than another, they weren't doing so on the basis of knowledge explicitly of underlying physical characteristics and the operation of general laws. Rather, they used their imaginative understanding of him to get a sense of how the action was being generated. It was their empathetic understanding of Brutus that provided knowledge of which mental process was operative here. In fact, our ordinary understanding of the concept of mental process seems to be provided by this capacity for imaginative understanding.

It's natural to think that this imaginative understanding that we have of one another must be epistemically somewhat shaky.

Surely a definitive knowledge of which motive someone acted from is available only through science. Our imaginative understanding of one another provides only speculation, but that line of thought leads nowhere. Scientific approaches are good for determining general causation in the mental and in the physical, and for finding singular causation in the physical. Finding singular causation in the mental is the domain of a distinctively humanistic understanding of one another.

And in fact, we do not usually regard our imaginative understanding of one another as merely speculative. We regard our ordinary knowledge of which motive someone acted from as capable of meeting the highest possible epistemic standards. In the law courts, and in everyday life, we think it can be known "beyond reasonable doubt" from which was the motive that someone acted. We also take it that this knowledge is sufficiently secure to ground putting someone to death, releasing them, or any of a range of serious outcomes. In practice, we regard our imaginative knowledge of one another's causal mental processes as meeting the highest possible epistemic standards, even when it isn't grounded in neuroscience—or indeed any other kind of science.

CHAPTER ONE

THE SPACE OF REASONS AND THE SPACE OF CAUSES

1. REASONS VS. CAUSES

Many people have thought there are differences between mental causation and physical causation. The differences have seemed so weighty that many philosophers have thought that mentalistic explanation ought not to be called "causal" at all. One way in the idea is put is by contrasting the "space of reasons," where we find properly psychological explanations of behavior, with the "space of causes" (Sellars 1956, §36; Rorty 1979, 157).

> In characterizing an episode or a state as that of *knowing,* we are not giving an empirical description of that episode or state; we are placing it in the logical space of reasons, of justifying and being able to justify what one says. (Sellars 1956, §36)

But on the face of it, there is no immediate tension between the existence of normative relations between psychological states and actions and the idea that psychological states can be the causes of actions. That is, we can recognize that there's a difference between the "space" of normative relations among psychological states and the "space" of causal relations among psychological states and actions without accepting that psychology provides only explanations in the space of reasons.

In fact, recognizing a role for normative considerations in psychological explanation seems to require that we think of psychological explanations as causal. After all, it is not as if the normative relations between the psychological states are thought to hold outside the ken of the subject. The natural thought is that it's because the subject recognizes the force of the normative relations that the psychological states propel the subject to action. The beliefs show the action to be a good idea, and it's the subject's recognition of this that causes there to be a causal relation between the states and the action. If the subject hadn't recognized that the beliefs made the action a good thing, the beliefs wouldn't have propelled the subject to action. This way of thinking of things seems to require that the subject was caused to act by the reasons. The normative relations provide the scaffolding within which these causal relations hold.

There's a particularly lucid version of the idea that psychological explanation is not causal explanation in Dennett (1981). On Dennett's picture, the relations between psychological states to which we appeal in characterizing someone's action are indeed normative relations. He talks about "the intentional stance" in which one uses the rational, normative relations between psychological states in characterizing the action (1981, 61). To talk in psychological terms—to adopt "the intentional stance"—is to suppose that the subject is rational. These normative relations, however, are not exploited to find the causes of the action. Rather, the use of psychological terms, the appeal to the normative relations found in the intentional stance, is not an attempt at causal explanation at all. Therefore we have here a very strong distinction between "the space of reasons," in which we find psychological explanations, and "the space of causes." What causes the action is stuff in your brain; the normative relations among propositions matter only for the exercise of predicting behavior, not for explaining it causally. The position is made vivid by an example from Somerset Maugham.

> There was a merchant in Baghdad who sent his servant to market to buy provisions and in a little while the servant came back, white and trembling, and said, Master, just now when I was in the market-place, I was jostled by a woman in the crowd and when I turned I saw it was Death that jostled me. She looked at me and made a threatening gesture; now, lend me your horse, and I will ride away from this city and avoid my fate. I will go to Samarra, and there Death will not find me. The merchant lent him the horse and the servant mounted it, and he dug his spurs in its flanks and as fast as the horse could gallop he went. Then the merchant went down to the market-place and he saw me standing in the crowd and he came to me and said, Why did you make a threatening gesture to my servant when you saw him this morning? That was not a threatening gesture, I said, it was only a start of surprise. I was astonished to see him in Baghdad, for I had an appointment with him tonight in Samarra. (Maugham 1933, Act 3, 112)

The key point here is that Death does not know the causes of things. Death does not know that it was the jostling that caused the servant to go to Samarra. Death has an appointment book (we may surmise) and can predict where people will be at particular moments. But the appointment book does not of itself offer any insight into why any one person will be at any place at any time. In Dennett's picture, being well versed in common-sense psychology is like having Death's appointment book. You have a way of generating predictions about who will be where when, but you do not as yet have any insight into why they behave as they do.

Dennett summed up his view like this. The "intentional stance"—what we use when we're talking about people's minds— requires an assumption of rationality on the part of the target. We assume they're kind of sensible. For example, "sheltered people tend to be ignorant; if you expose someone to something he comes to know all about it . . . our threshold for accepting abnormal ignorance in the face of exposure is quite high. 'I didn't know the gun was loaded,' said by one who was observed to be present, sighted, and awake during the loading, meets with a variety of utter skepticism that only the most outlandish supporting tale could overwhelm" (Dennett 1981, 61–62). The "intentional stance" is like Death's appointment book. It allows you to predict what someone is going to do without giving any insight into causes. Being predictable in this way is all it takes to have a mind, because the only use for our mentalistic talk is in giving these kinds of prediction:

> Any object—or as I shall say, any system—whose behavior is
> well predicted by this strategy is in the fullest sense of the
> word a believer. What it is to be a true believer is to be an
> intentional system, a system whose behavior is reliably and

> voluminously predictable via the intentional strategy. (Den-
> nett 1981, 59)

This is a version of the idea that the "space of reasons" in which we find psychological explanation is quite different to the "space of causes." When we are operating in terms of the space of reasons, we are merely using a predictive abacus. Finding causes would require us not to be talking in mentalistic terms at all but rather to be looking at the biology of the brain.

Dennett does not give much weight to the idea that the normative connections that characterize the space of reasons are known about in a way that is quite different to the way in which we know about the causes of things. But for many philosophers, the trouble with supposing that mentalistic connections are causal connections comes when we reflect on the normative relations that characterize the "space of reasons." The idea is that these normative relations must themselves be a priori, or, as philosophers used to say, in some broad sense, "logical." But when a priori or logical connections hold between two states, it's said, they can't be "distinct existences" in Hume's sense (Hume 1748 / 1975, IV / 1). But cause and effect have to be distinct existences. Therefore, the relations between these normatively linked states can't be causal. Moreover, traditionally there is no predictive point to the exercise. It is usually thought to be just one of the fundamental modes of explanation, when one shows why what someone did is normatively correct in the light of one's mental states. There may be no particular predictive payoff from the exercise. Indeed, it's often thought that common-sense psychology is not predictive. Consider, for example, an ordinary conversation with someone you know well. You might be quite unable to predict what they're going to say next; indeed, that's one element that makes ordinary conversation worthwhile. But that inability to predict doesn't of itself mean that it will be incomprehensible to you why the other person says the things she does. In fact, every single thing she says might be perfectly explicable in terms of her mental life, in that you understand why she says every single thing she does and why it was a good thing to say given the rest of her mental life.

This whole line of thought of separating the space of reasons from the space of causes is quite wrong. We can and do give causal mentalistic explanations, and the notion of causation here is exactly the same as the notion of causation that we use in the physical case. And although normative considerations do have a role in psychological explanations, they have a quite different role than is here envisaged.

2. RANDOMIZED CONTROLLED TRIALS OF THE MIND

Suppose you have to assess a number of paintings for their pictorial merit. Without being professional critics or even knowledgeable amateurs of art, many people would be willing to do this. You might have to do it without knowing who the painters are, or you might be told the painters; some of them may be quite famous. Would knowing who the painter was affect your assessment of pictorial merit? Judgments of specifically pictorial merit presumably ought to be independent of knowing the painter. But in a study by Hansen et al (2014), most of their subjects thought that the judgments of other people would be biased by knowing who painted what picture, and they thought their own judgments would be similarly biased. For the most part, each of the subjects thought that they themselves would tend to give more favorable assessments to the pictures known to be by famous painters. The subjects in this experiment were then divided into two groups. In one group, each person was given an array of paintings to assess for pictorial merit but without any identification of the painters. The other group was given the same array of paintings to assess, but some of the paintings were labeled with a famous painter as the author. The subjects who had been told the painters of various pictures did their best to assess the painting on their purely pictorial merit. In fact, they thought they had managed to do this. They thought their assessments were independent of the knowledge. But comparing their assessments to those of the group that had been given no information about the authors of the paintings clearly showed the effect. They were rating more favorably the pictures that they knew to be by famous painters.

This study illustrates a number of points about implicit bias. Most strikingly, it shows that

a. a belief, in this case a belief that a painting is by a famous painter, can be causally impacting your assessment of its pictorial merit, even though

b. you sincerely think that this belief is having no impact on your judgment, and

c. this can be so even though you agree that this kind of belief does in general have a biasing effect and even though you have agreed that you yourself are likely to be subject to this effect.

The whole point about these findings is that they're established by scientific study. They aren't manifest to common sense. You might suspect that they're true or that they aren't, but it's a scientific study that determines whether they're true.

On the face of it, this kind of study demonstrates that the dynamics of the mind can be studied by science in very much the same kind of way as the dynamics of any physical system. We can investigate the causal impact of a belief—in this case, the belief that a painting is by a famous painter—on your other thoughts and feelings. We do this using exactly the same experimental methods science uses to establish the causal impact of physical factors. In particular, we are here using a randomized controlled trial. To give a physical example, suppose you're trying to determine whether a drug has any effect on an illness. You divide your subjects into two cohorts. You do this "randomly" in the sense that there isn't any systemic factor differentiating the two groups that might conceivably have any causal bearing on the illness. You give the drug to people in one of your cohorts but not to the people in your other cohort—this second group being your control group. That's the sense in which the trial is "controlled." Then you look at whether there is any difference, on average, in the incidence or severity of the disease across the two groups. If there is any systematic difference between the two groups, that can only be an effect of the drug.

The logic here is compelling. Each of us trusts it whenever we take a drug. Of course, in any particular case, there are many problems to discuss about whether the general idea has been correctly implemented. For example, it's notable that "random" here refers to the outcome of the process of selection rather than to the process of selection itself. Suppose people were sorted into two groups by flipping a fair coin for each of them to determine into which group they'd go. In one sense, the process itself then would be "random": the factor used to determine which group a subject goes into, the flipping of the coin, presumably has no systematic causal impact on the outcome we are interested in (the disease the drug might protect against, for example). But the point of the thing might still not have been achieved. It might have been that by accident, one group had all the people with high levels of a hormone that confers natural resistance to the disease, and the other group had low levels of that hormone. In that sense, we would not have a successful randomization. We try to minimize the likelihood of this happening by having large group sizes. But it is often possible to wonder, in particular cases, whether group sizes have been sufficiently large and whether sufficient measures have been taken to guarantee that factors relevant to the outcome have not been operating to affect which group a patient gets into. For example, socioeconomic status might be a factor that in one way or another affects the incidence or severity of the disease. It might be that in the case of a disease for which there is no known drug treatment, patients of high socioeconomic status are particularly good at getting themselves into the treatment groups for experimental drug therapies. Still, these are problems in applying the general design. They do not point to any difficulty with the general design itself. This is the key way in which we experimentally demonstrate causality.

The very same design is what we use to demonstrate causality in the mind. That's what was used in the example of knowledge about painters above. We divided our subjects into two groups, one of which got the information about painters and the other of which did not. To vary the example a little, suppose that we want to find the causal impact of a belief that the applicant for an academic post is a woman on the evaluations made by academic assessors. We divide our assessors into two groups. We give the two groups exactly the same information about the candidate, except that that one is given the information that the candidate is a woman and the other is not. Any systemic difference between the evaluations made by the two groups can be caused only by the difference in whether they believe that the candidate is a woman. That's the logic of the design. Of course, here as in the drug trial, it is often possible to argue about whether the design has been correctly implemented. You can argue about whether we have managed to randomize the two groups in the sense that there isn't any prior systemic difference between them, and so on. But the design itself seems compelling. We have here a way of establishing causation in the mind.

Indeed, arguably all of us use this way of establishing causation, at least implicitly, from an early age. Children show a sensitivity to statistics, and a use of statistics in causal reasoning, from astonishingly young ages (for reviews, see Gopnik and Wellman 2012 and Xu and Kushnir 2012). A sensitivity to statistical patterns shows up already at eight months old. For example, Xu and Garcia (2008) set up an experiment that used the fact that infants generally look longer at unexpected events. There were two urns. One visibly contained mostly white balls. The other visibly contained mostly red balls. When the researcher took a sample of mostly red balls from the mostly white urn, infants looked longer at it than they did at a similar sample of mostly red balls taken from the mostly red urn. They implicitly recognized that the probability of a mostly red sample being drawn from the mostly white urn was less than the probability of a mostly red sample being drawn from the mostly red urn. Kushnir, Xu and Wellman (2010) spun this point in an experiment with 20-month olds. The researcher again had two urns, one of which contained mostly toy rubber ducks. The other urn held mostly toy rubber frogs. Similarly to the Xu and Garcia paradigm, there were two conditions. In one, the researcher took a handful of all toy rubber frogs from the urn holding mostly ducks and played enthusiastically with them. In the other condition, the researcher took a handful of toy rubber frogs from the urn containing mostly frogs and played enthusiastically with them. The behavioral cues to emotional preference were the same in the two conditions. But the children in the first condition were more likely to select a frog to give to the researcher than the children in the second group. This indicates that children were using the nonrandom sampling as a guide to preference; the unlikely selection of all frogs from the mostly ducks urn was being taken to reveal a preference for frogs, whereas selecting only frogs from the mostly frogs urn was not interpreted as exhibiting any particular preference for frogs.

Alison Gopnik and her colleagues have argued for some time that children from around age four or five will use experiment and observation to find which characteristics of an object will make a machine's lights flash or sound a bell, for example (Gopnik et al. 2004). They try objects with the characteristic and objects without the characteristic to see whether it makes a difference to the outcome. They also seem to be capable of using this approach to human psychology. Betty Repacholi and her colleagues set up an experiment in which children were presented with a bowl of cookies and a bowl of broccoli, and an experimenter asked them to give her food from one or other of the bowls. The experimenter enthusiastically mimed distaste for the cookies and a strong liking for broccoli. Nonetheless, children at fourteen months resolutely fed her cookies. They seemed not to understand the pos-

sibility that she might prefer broccoli. At eighteen months, however, the children transitioned: they fed the experimenter broccoli.
Repacholi hypothesized that children only begin to understand
the very possibility of divergent desires around eighteen months
old. She pointed out that this understanding precedes the onset of
the "terrible twos," the period of teasing and messing about children around the age of two delight in: "at around 18 months
children begin to experiment with these desire conflicts, often
intentionally setting up conflicts of desires and observing the results (the typical behavior of 'the terrible 2s'). This may also suggest that these children are constructing a theoretical understanding of desire as a way of explaining apparently confusing evidence
about human behavior" (Repacholi and Gopnik 1997, 19).

Indeed, arguably we carry on using these ways of finding out
about the causal implications of the psychological all through our
adult lives. We learn and achieve better understanding of the sub-
tleties of courtesy, for example, and how varying levels of courtesy can impact social interactions. We get to understand how
observance of courtesy can trade off against authenticity of expression in dealing with other people, for example. We learn better what will work and what won't in conveying sympathy, and
so on. Perhaps much of this learning is at an implicit level, affecting the ways in which we interact with other people without
necessarily being something that we could make explicit. But
whatever level it's at, this kind of learning about psychological
causation seems really basic to ordinary social life.

This ordinary, common-sense causal learning uses the same
methods as physical science, and it seems to take us a long way. It
also seems, though, that an explicitly scientific approach could

take us a great deal further in understanding the dynamics of the mind. In the case of implicit bias, for example, it seems evident that explicitly designed scientific studies can provide a broader and fuller understanding of what is causing what than we could achieve at the level of an individual using their own trial-and-error experience to determine what makes what happen. There doesn't seem to be a way in which, without some scientific or semiscientific study, you could demonstrate implicit bias in academic hiring, for example, particularly given how invisible implicit bias can be to all of us affected by it. Our ordinary, common-sense understanding of what causes what in the mind seems to be open to justification or criticism by explicitly scientific studies.

The possibility of pursuing a scientific approach to causation in the mind, as the example of implicit bias shows, demonstrates that the psychological life exhibits a causal structure in the same way that a physical system can exhibit a causal structure. It's hard to sustain the view that there are any general characteristics of the mind that will resist this kind of study by science. It's not as if there is any domain of the mind where causation can't be probed scientifically. Let's end this section by giving some ways in which the reach of science can be dramatically extended.

It's true that there are some things that are quite special about the case of implicit bias. For example, in giving two sets of subjects two sets of applications for a position, differing only in that one set of applications identifies the candidates as women, we can be fairly confident that the two sets of subjects differ relevantly only in their beliefs about whether the candidate is a woman. But suppose that we wonder whether poverty is a cause of mental illness. As you go to the bad parts of town, the incidence of mental illness increases. But what is the direction of causation? Is there a genetic or environmental common cause of poverty and mental illness? Or does the mental illness cause the poverty? These questions might seem to resist empirical inquiry because we can't go around inflicting poverty on one group of subjects and withholding from another simply to find out what it does to you. How could you devise an experiment to determine whether poverty causes mental illness? There are limitations on how we can investigate human subjects that might seem to be limitations on the scientific method as such. But the use of the experimental scientific approach to psychological causation does not demand that we should be able to manipulate at will the characteristics in whose causal significance we are interested. Beginning in 1993, Jane Costello and her colleagues on the Great Smoky Mountains Study of Youth recruited 1,420 children aged nine to thirteen years old from western North Carolina to an eight-year study of the development of psychiatric disorders. In the study, 350 of them were American Indian children living on a reservation in the study area. Quite coincidentally, three years later, a casino opened on the reservation, and from then on everyone on the reservation, adult or child, received a percentage of the profits. So here we have a "natural experiment." The opening of the casino wasn't any investigator's plan, yet the amount paid out was enough to lift many of the affected families out of poverty. This was a "natural" intervention on poverty. Suppose there is a difference in the incidence of mental illness among those who have been lifted out of poverty and those who have not been lifted out of poverty. That must reflect a causal connection from poverty to mental illness. Costello et al. (2003) found that there was indeed a correlation between mental illness and being lifted out of poverty, confirming that poverty is a cause of mental illness.

It is worth briefly remarking on the depth and subtlety that is possible in studies that derive psychological causality from statistical associations, even in cases where an intervention by the researcher is not possible. For example, it has long been known that there is an association between stressful life events and major depression (e.g., Surtees et al. 1986). To find whether the link is causal is not straightforward. In the case of implicit bias, we can take our subjects, divide them into two cohorts, and administer the belief that a picture is by a famous painter, or that the candidate is a woman, to one cohort but not to the other. But we cannot divide our population into two cohorts and administer stressful life events to one but not the other. Moreover, we can see how the correlation might be explained without presuming a causal link. Some people may have a genetic tendency to put themselves into situations where they are likely to experience stressful life events, and those genes may be correlated with genetic risk factors for major depression (Kendler and Karkowski-Shuman 1997). In response to this point, we can divide stressful life events into two: those that seem to be dependent on the behavior of the subject, and those that do not. Correlations between independent stressful life events and major depression are more likely to be directly causal rather than being products of an underlying genetic correlation. Kendler, Karkowski, and Prescott (1999) looked at the relationship between independent stressful life events and major depression, particularly within monozygotic twin pairs. Monozygotic twin pairs share their genetic and familial-environmental backgrounds. Therefore we can, in effect, regard the infliction of an independent stressful life event on one of the twins but not the other as a natural intervention, for which the other twin is the control. And in fact they found that those independent stressful life events within twin pairs were strongly associated with the onset of major depression. Then in a further twin study, Kendler et al. (2003) looked at which aspects of stressful life events were responsible for major depression. They classified life events using dimensions of loss, humiliation, entrapment, and danger. They found that loss and humiliation were particularly significant for major depression; entrapment and danger had no significance for major depression. (For many further uses of "natural experiments" in psychiatry, see Rutter 2007.)

Finally, consider the question whether drug addiction is voluntary, under the subject's control, or to be assimilated to a physical disease, for which the subject is not to be held responsible, except insofar as they were responsible for their initial exposure to the risk factors for contracting the disease (e.g., Leshner 1997). At first this seems an extremely difficult problem to investigate scientifically, but we can chip away at it. In a recent study, Kendler et al. (2017) looked at the registries across Sweden in a population-wide study of the causal impact of pregnancy on substance use. They found that pregnancy had a significant impact on substance use. For example, using a model that fitted the data well, "we could predict a reduction in risk for drug abuse of 83% in a pregnant woman compared to her non-pregnant monozygotic co-twin." Now, this is not a true intervention, and you might argue that perhaps the pregnancy has a direct biological impact on the brain's need for, say, cocaine. But it seems more immediately plausible that what is happening to reduce cocaine use is concern for the child. This suggests that a high level of motivation can impact substance use. Systematic understanding of the psychological causal factors underpinning addiction and how they relate to biological factors is by no means impossible, though it is difficult.

Of course, psychiatry and the social sciences face difficult problems in trying to construct experiments or to observe natural experiments. For example, people have talked about the "replicability crisis" in psychology and the biomedical sciences, where it has proven impossible to replicate many important studies (e.g., Benessia et al. 2016). This issue is of considerable practical and theoretical interest, but it does not affect the point of the present section. First, the issue does not threaten the possibility, in principle, of deriving causal links from statistical evidence, and in particular statistical evidence about what happens under interventions. People have proposed that the problem should be addressed by having much more rigorous conditions on the description of how a study is set up so that the thing cannot be published without, in effect, instructions on how to replicate it or cautions against the mechanical application of statistical programs to data (Stark 2018; Stark and Saltelli 2018). But none of these responses challenge the very idea of deriving causal information from statistical evidence about what happens under interventions. Moreover, the problems here are by no means unique to the psychological sciences. It is not easy to quantify these things, but the situation seems to be similar and just as bad in biomedicine, for example. There are some problems, such as knowing when one has a valid psychological construct to measure (e.g., happiness or intelligence) that seem to be peculiarly difficult in the sciences of the mind. But most sciences do struggle with this kind of issue (see Chang 2004 on the problems physicists had in finding a well-grounded conception of temperature). Or consider the difficulty of finding a population on which you can ethically conduct an experiment to find whether cannabis use causes schizophrenia, or the difficulty of replicating a study when people in one part of a country can vary so unpredictably from people in another part of the same country, let alone another country. These problems are also recognizable in the science of climate change, where the significant variables are not psychological at all. It's practically difficult and often ethically impossible to conduct legitimate experiments, and regions vary so widely in so many parameters that replications are often impossible in practice. But in principle, these kinds of problem don't indicate anything mistaken about the idea of studying causal relations among the variables involved in climate change, and neither do they of themselves indicate anything wrong with the idea of studying causal relations among psychological variables (cf. Knutti 2008 for more raw data about the problems involved in climate change; Glymour 2007 finds a parallel between weather and the psychology of the individual).

Indeed, for all the intense focus we have put on understanding one another psychologically for at least the last few thousand years, it seems conceivable that we are only beginning on the possibilities of this kind of research into how human psychology works. The possibilities now of harvesting enormous quantities of information about people's preferences and decisions, and processing that to find causal implications for how human psychology works, seem at the moment to be limitless and likely to yield significant further insights—and dangers.

3. DAVIDSON: THE LOGICAL CONNECTION ARGUMENT APPLIED TO SINGULAR CAUSATION

On the perspective that we are finding so far, the status of psychological factors as causes is one thing, and whether they rationally justify an outcome is another. Let's suppose we find that the belief that one is pregnant causes a reduction in substance use. Because that belief is causal, it can be part of the explanation of a reduction in substance use. Whether the belief rationally justifies a reduction in substance use is a further question. Of course, there is the case in which people review the matter and conclude that the belief would justify a reduction in one's substance use, and having come to that conclusion that the thing is rational might in turn cause them to allow the causal action of the belief in reducing substance use. But that's not always what happens, and the Kendler et al. (2017) result does not depend on supposing that this is ever what happens. Or consider that a belief that someone is a member of a particular ethnic minority might cause the people in a particular community to suppose that person is likely to be armed and dangerous. That can be so even if everyone has explicitly reviewed the situation and concluded that the belief that people are of that minority doesn't rationally justify supposing them to be armed and dangerous. We have the patchwork of psychological states whose causes and effects we study by experiment and observation, and we have the question of what rationally justifies what. These are different questions, even though one might speculate that there may be evolutionary reasons to suppose that humans probably do achieve some kind of approximation to rationality.

Let's look at a different analysis of the distinction between the space of reasons and the space of causes. The argument to which Davidson was responding in his classic article on reasons as causes (1980) was the "logical connection" argument. The idea was that the reasons we appeal to in explaining an action have a normative, a priori or "logical" link to the outcome, and therefore can't be properly distinct from the outcome and can't be causes of the outcome.

> According to Melden, a cause must be "logically distinct from the alleged effect" (52); but a reason for an action is not logically distinct from the action; therefore, reasons are not causes of actions. (Davidson 1980, 13)

For example, "to intend to do X" is simply to be such that, all other things being equal, one will do X. There's a logical connection between the intention and the action. Therefore, the intention can't be a cause of the action. As Davidson said, one way we can make this argument more fully explicit is to appeal to Hume's idea that cause and effect must be "distinct existences."

> One possible form of this argument . . . Since a reason makes an action intelligible by redescribing it, we do not have two events, but only one under different descriptions. Causal relations, however, demand distinct events. (Davidson 1980, 13–14)

Davidson then gave two lines of response to the argument.

1. We can have logical relations between the descriptions of two distinct events. So long as the events are distinct, one may still be the cause of the other. For example, "the cause of B caused B" may be true, even though the descriptions identifying the two events ("the cause of B" and B) are logically related (Davidson 1980, 14).
2. "Desires cannot be defined in terms of the actions they rationalize, even though the relation between desire and

action is not simply empirical; there are other, equally
essential criteria for desires—their expression in feelings
and in actions that they do not rationalize, for example"
(Davidson 1980, 15).

I think the first point to make about this strategy is that it leaves
us with no analysis of statements such as "humiliation causes depression" or "belief that one is pregnant causes cessation of smoking." Davidson is talking only about singular causation, a relation
between particular concrete events (or episodes, or trope instantiations; there is some variation in how people think the individual concrete items here should be characterized, but that does not
matter for the main points here). General causation, in contrast, is
a relation between variables. In the previous section, we were
looking at the kinds of randomized controlled trials that establish
general causal truths, such as "humiliation causes depression."
But Davidson's points here apply only to singular causation. So
far as his response to the "logical connection" argument goes, the
relation between desire and action is "not simply empirical" (1980,
15), so we are left with the idea that "desire causes action" cannot
be a correct causal claim because desire and action are not distinct
existences. But again, there seems to be no reason to accept this;
we can again divide a group of subjects into two cohorts, supply
(in any of the ways suggested above) a desire to one group but not
to the other, and see whether there is any systemic difference at
the level of action.

Argument (1) leaves dangling the question how to address the
Logical Connection Argument as it applies to general causation.
One possibility, left open by Davidson's point, is that the Logical
Connection Argument is correct for general causation. That is,

you might acknowledge that the existence of grammatical or logical connections between psychological variables shows that general causal claims stated using variables, such as "intention causes action," usually merely reflect logical relations rather than true causal connections. Argument (2) equally leaves this possibility open. Suppose we accept that we can't define desires in terms of actions. Still, because the relation between desire and action is "not simply empirical," what does explain the existence of correlations between desire and action? It will presumably be the nonempirical connections between the two rather than any causal relation. But as we'll see, there is really no reason to believe in these nonempirical connections. Desire is one thing, and action is another.

Davidson's problem was that he had two real insights: that psychological explanation is causal explanation, and that rationality is somehow involved in psychological explanation. But his view of causation left him without any way of reconciling those points. Suppose we acknowledge that psychological explanation is causal explanation. Insofar as it's causal explanation, we have to acknowledge that in principle, anything can cause anything—in particular, that anything psychological can cause anything else psychological—and the methodology of the social sciences and psychiatry has no a priori investment in the presumed rationality of humans. Because Davidson viewed causation as grounded in exceptionless laws, the only way he could find a constitutive place for rationality in causal psychological explanation would have been as a condition on alleged psychological laws governing behavior—that we can fill out a background pattern of psychological laws that articulate what rationality is and to which the human mind conforms. But as he pointed out, there are no such laws to be formulated

(Davidson 1980, 233). That meant that rationality, and the psychological in general, seemed to be left with no role in causal explanation, and it left Davidson open to the charge that on his view, the mental was epiphenomenal (e.g., Honderich 1982). So how should we hold on to the idea that psychological terms can figure in causal explanations? The most popular approach (Jackson and Pettit 1990; Yablo 1992) was to say that there may be counterfactual dependencies between a psychological characteristic and a behavioral outcome. But that approach leaves us without any special place for rationality because counterfactual dependencies can in principle hold between any collection of psychological characteristics and any behavioral outcome. In the next chapter, we'll see that there is indeed a role for rationality in singular causal psychological explanation, but that we can find its place only by outlining the role of the concept of a process in an analysis of causation and locating types of mental process.

The correct reaction to the above points about general causation is surely to accept that there are normative relations between psychological states, and these normative relations may be a priori, but that there is no a priori presumption that the causal dynamics of human beings will reflect these normative connections. The methodology of randomized controlled trials for finding the causal relations among psychological variables in human populations makes no presumption of rationality. When we divide our population into two randomly assorted cohorts, systematically differentiated by only a single psychological state, and look for any systematic difference between them, there is no presumption that the outcome will be rationally justified. For example, when we look to see whether the belief that a painting is by a famous painter has any causal impact on one's assessment of its pictorial merit, that is quite independent of any question about whether the belief that a painting is by a famous painter rationalizes a high view of its merit. You might think it is rational and you might not; that's strictly irrelevant to the outcome of the experiment and the determination of causation. Using this methodology, you might find of any belief that it's the cause of headaches, or the belief that dragons are extinct, or really just anything at all. Similarly, when using randomized controlled trials to find the effect of a drug, there isn't any a priori presumption that the result is going to go in one direction or another.

This is why there is such a discordance between social psychologists and philosophers on the subject of rationality (Thagard and Nisbett 1983). Philosophers generally take it that there is some kind of presumption of rationality in psychological explanation: that the point of psychological explanation is to display the rationality of the subject. When Tversky and Kahneman first came out with their studies exhibiting human irrationality (1981), they were using an experimental methodology that had no commitment to finding rationality on the part of their subjects. Philosophers took it that there must be something wrong with Kahneman and Tversky's results. Philosopher L. Jonathan Cohen, for example, said that there must be a place for a conception of "rational competence" possessed by all humans with minds, such that this cognitive competence "corresponds point by point with the normative theory" (Cohen 1981, 321). To psychologists without an a priori investment in human rationality, such remarks merely reveal the sterility of a philosophical tradition not properly informed by empirical work. Of course, it has been possible to review work like Tversky and Kahneman's and ask whether it really does demonstrate failures of human rationality, rather than the compromises that have to be made when optimizing the use of one's bounded cognitive resources: it's one thing to be trying to solve a puzzle when one has three days to focus on nothing but it, and it's another thing when one has only a few seconds and the thing is of little interest anyway. Such an approach relocates the role of rationality from being an a priori constraint to being something that may be expected to have arisen naturally as a product of evolution. The simplest version of such a view is Jerry Fodor's atomism (1998), on which beliefs are ascribed one by one: a belief is simply a neural state nomically locked on to some external state of affairs. All the beliefs can be ascribed independently of one another. Maybe there will be some evolutionary advantage to having the organism turn out to be broadly rational, but that's all there is to the demand for rationality: it has no constitutive force. Even if we acknowledge a place for some kind of holism in the ascription of attitudes, it is unlikely to vindicate a global rationality, at least as "rationality" has been classically understood. Humans, like other animals, are operating with limited time and limited cognitive resources, and they have to optimize their use of them. The best we can hope for is some notion of what Thomas Griffiths et al. have called "resource rationality" (2015), in which the subjects do the best they can given the brain they have. Of course, there is a threat of vacuity in this approach: perhaps anything the subject does in reacting to a situation can be represented as them "doing the best they can, given the brain they have." And the constraint imposed on what a subject can rationally think in this sense of "resource rationality" is minimal. Even the delusions of the schizophrenic can be represented as rational. As Lisa Bortolotti (2016) points out, there are epistemic benefits of succumbing to delusion. For the prodromal schizophrenic patient, there is a high level of anxiety, trepidation, and preparedness for crisis preceding the onset of delusion, which makes ordinary cognitive learning and ordinary adjustments to one's surroundings difficult. Once we have the full onset of delusions, however, there can be a certain relief from anxiety in it that makes much of ordinary cognition possible again, albeit under the sway of the delusion.

Davidson's idea that human reasoning must be rational was a development of Quine's line of thought: that a finding of irrationality can always be replaced by a finding of mistranslation (Quine 1960, 58–59). Quine's idea is perhaps best illustrated by his remarks on "the myth of a pre-logical people." Suppose, Quine said, an anthropologist says that there are people who reason as follows: to prove "if p then q," they demand that p should actually have been shown to be true and that q also should have been shown to be true. Moreover, suppose that in addition to this display of irrationality, they should also make the following crazy move: when drawing implications from "if p then q," alone, they allow p to be drawn as a conclusion, and they also allow q to be drawn as a conclusion. Well, the anthropologist says, marvel at the condition of these people who badly fumble conditional reasoning. Even the youngest children in the anthropologist's own community do not make such elementary mistakes in reasoning. But, Quine points out, there is an alternative analysis available: perhaps we have mistranslated, and what we are taking to be the conditional is actually a phrase structure for conjunction. Quine and Davidson said the general case is that we can always substitute a finding of mistranslation for a finding of irrationality. The trouble is that the general claim does not seem to be true. It would, for example, be a bizarre reaction to a social-psychology study showing how bad people are at gauging the probabilities in a game of poker to suggest that those people are speaking a deviant form of English. And the point seems to be written large in cases of psychosis. Someone who believes that his body has been taken over by a lizard cannot plausibly be said to be rational (cf. Browning and Jones 1988). Of course, if you are committed to thinking of human psychology as through and through rational, you may say that even in this kind of case, there is rationality. After all, if you did believe that your body had been taken over by a lizard, wouldn't you behave as this person does? But you can only do so much moving around the bump in the carpet. At a certain point, you have to ask, "Where does the belief that your body has been taken over by a lizard come from?" That is not itself a rational matter, but the ordinary English words give the content of the delusion.

4. INTERVENTIONS

Let's try to set out more explicitly the idea underlying the use of experiments to establish causation. The idea scientists use in demonstrating causation is that there's a connection between causation and what happens when there's an intervention on a system. In conversation, you will often find scientists talking as if the question whether A causes B is a notational variant of the question: if an experiment were to manipulate A (in ideal experimental conditions), would there be a difference to B? It's one thing to notice that people who are having hormone replacement therapy are generally in better physical condition than people who are not. But is it the hormone replacement therapy that is responsible for the benefit to their physical condition? The way you find this out is by doing an experiment. Divide your subjects into two cohorts and give one but not the other the hormone replacement.

Then any systemic difference in physical condition between the members of your two cohorts must be due to the therapy.

What is an experimental intervention? The basic distinction we need here is between the variables characterizing the ordinary causal functioning of a system—such as a human being, or an economy—and those that are exogenous to the system. An intervention is when a variable external to the system comes "from outside" and seizes control of some element of the system. Suppose we have a large number of variables characterizing the ordinary functioning of some complex system, such as the human body. We find a correlation between two of the variables, X and Y—for example, salt intake and blood pressure. We wonder whether there is a causal connection between them. Is the salt intake a cause of blood pressure? We can't simply read off from the correlation that there is a

Figure 1.1. We assume as our starting point that we have a set of variables characterizing the ordinary function of a (possibly very complex) system such as an economy, the human mind, or the human body, and a correlation between two of those endogenous variables, *X* and *Y*.

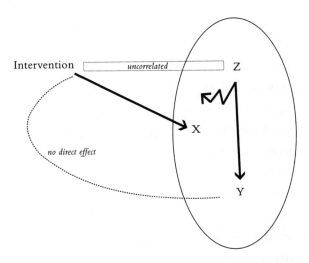

Figure 1.2. This represents an intervention as the action of an external variable on one of a set of variables characterizing the ordinary functioning of a system (there may in any particular case be many more variables than X, Y, and Z characterizing ordinary functioning). The intervention variable has to seize control of the target variable X, suspending the influence of any of its ordinary causes Z that are also influencing the outcome Y. The intervention itself mustn't be correlated with any variable Z that also affects the outcome Y. And the intervention mustn't affect the outcome Y directly— that is, otherwise than by affecting X. (This diagrams the analysis of Woodward 2003 and Woodward and Hitchcock 2003.)

causal connection. The fundamental problem is that there may be some third factor, Z, that is causing both the level of salt intake and the blood pressure. To find whether X truly is a cause of Y, whether salt intake is a cause of blood pressure, we do an experiment. We come from outside the system, the experimenter seizes control of salt intake, and we look for systemic differences in blood pressure depending on salt intake. If we do the experiment correctly, we establish whether salt intake is a cause of blood pressure (see figs. 1.1 and 1.2).

So far, I've given an abstract characterization of what an intervention is, following the lines of Woodward (2003) and Woodward and Hitchcock (2003). I think that this characterization is good for bringing out the most abstract connection between interventions and causality. But to look further at the parallels and contrasts between causation in the mental and causation in the physical, we will have to distinguish between three types of intervention. All of them fall under the abstract characterization, but there are significant differences between them.

1. The first grade of intervention is what we do in a randomized controlled trial, or in some types of "natural experiment." In this kind of study, individuals are shuffled into cohorts so as to randomize possible confounding factors. In the simplest cases, we have two cohorts, and we give the treatment to one cohort but not to other. If there is a difference in the incidence of the outcome between the two groups, then we conclude that the treatment caused the outcome.

2. The second grade of intervention is what we do in an experiment where we have selected two individuals so that we know them to be alike in all possible confounding factors, and then we apply our treatment to one of them rather than the other. This is the common approach in molecular and cell biology: we deal with two sets of cells grown in the same culture and apply our treatment to just one of them; any difference between the two sets of cells then must reflect the causal action of the treatment.

3. The third grade of intervention is when we take a single individual and change just one characteristic of that

individual to see what difference there then is in the other characteristics of the individual. This is the natural behavior of a child with a new toy, or a guest in a new hotel room trying to find out what all the switches do. You flip the switch from off to on and back again to see what it does.

The three grades differ in how much focus they put on the individual subject when setting up an intervention. In the case of a randomized controlled trial, very little attention is paid to the idiosyncrasies of any one individual. Suppose you are running a randomized control trial to find whether tamoxifen protects against breast cancer. You may have tens of thousands of subjects in the trial, and getting a definitive result does not depend on knowing about any of them beyond whether they got the treatment and whether they got breast cancer. The randomization itself is enough to make sure that there is no third factor that is a cause of both tamoxifen ingestion and protection from cancer.

At our second grade, the controlled experiment with just two samples, we have set up two individuals so that there is absolutely nothing different between them in the way of potential confounding factors. Thus, when we change just one characteristic of one of them, any subsequent difference between them can be traced to that characteristic. The conclusion here may be in the first instance that the change in that individual caused a particular subsequent effect. But there may be a general conclusion about the effects of that property, though just how general will depend on the specifics of the particular case (e.g., which other features of the individual are mediating the consequences of the change).

At our third grade, the individual is in effect acting as its own control. Because we reached in to manipulate just one characteristic of it, leaving everything else unchanged, any further subsequent changes must be the result of that manipulation. And again, it may be possible to find a certain generality in the conclusion.

In cases involving psychological variables, we may use interventions of any of these three types, though it is always possible to question whether what happened in any particular experiment has really met the ideal conditions of an intervention. In attempting to do a first-grade intervention to find the impact of a teaching technique, for example, it is very difficult to be sure that one has sorted the subjects into cohorts where the only relevant variable differentiating them is whether the teaching technique was used. If a new way of teaching mathematics is trialed in New York and is a great success there, and then it is tried in Atlanta, is anyone surprised if it doesn't replicate? And even within a New York trial, ensuring that there's no systematic difference other than use of the new teaching method between the treatment group and the control group is manifestly difficult. The second type of intervention is, again, difficult to implement with humans. It's hard to be sure that we have two people who are really the same in all respects so that we can apply our treatment to one but not the other so differences between the two can be definitively traced to the intervention. Similarly, for the third type of intervention, people very often learn from what's happened, and there's no guarantee that you are dealing with the same kind of causal system after an intervention has been applied to an individual. Despite these difficulties, we can and do use the method of interventions to discover causation in the mental. Indeed, social psychology, and arguably psychology generally, depends on the uses of these methods. Each of us, in our own individual learning about other people, arguably uses these methods over the lifespan. These same diffi-culties also pervade our attempts to use experiments to discover causal relations between physical variables.

It's natural to speculate, as an interventionist theory of causa-tion does, that causation is merely a summary of "behavior under interventions" (Pearl 2000), or as Woodward (2003) puts it in by far the philosophically most fully elaborated account, that causa-tion is to be explained in counterfactual terms as a matter of what would happen under interventions. For X to be a cause of Y, in these terms, is a matter of X and Y being correlated under (po-tential) interventions on X. The notion of an intervention is often not explained by examples but in the abstract terms I used above. If we ask for an example of interventions in practice, perhaps the natural candidate is the randomized controlled trial. Random-ized controlled trials play a peculiarly central role in the literature on causation. They're usually taken to provide a way of finding out about causation that is in some sense canonical. The critique of an experiment is a matter of finding how closely it approxi-mates to an ideal randomized controlled trial.

There are a number of reasons why we should not take this analysis of causation just as it stands, giving the prominence that it does to ideal interventions such as randomized controlled trials.

1. We cannot generally take the point of a randomized controlled trial as being merely to find out what happens in the circumstances of a randomized controlled trial. Suppose that we are testing whether a drug makes a difference to headaches. If we find that it does, the drug may be licensed and released into the wild. We suppose that its causal

efficacy will continue even outside the context of ideal interventions, even if it's always taken, for example, in the context of other potentially confounding factors such as trying to get some quiet, lying down, or quarrelling with other people. The use that we will make of the discovery of a causal relationship goes well beyond what happens in an ideal intervention.

2. Similarly, even in the context of discovery, rather than action on the basis of a causal relationship, we often suppose that we can operate on the basis of "imperfect evidence," or evidence that does not at all involve us managing to set up an ideal intervention. For example, suppose that you're interested in the question whether cannabis legalization causes an increase in the use of cannabis (rather than, for example, merely an increase in reported cannabis use). You will not be in a position to set up a perfect intervention here. Nonetheless, we do take it in practice that there can be reasonable inferences here.

An analogy may be helpful. Suppose you define "solubility" as a matter of a substance dissolving when put in pure water. Pure water may be hard to come by. Maybe there isn't any around—it's all muddy or has some kind of chemical substance in it. Nonetheless, even in these circumstances, you could provide a reasonable basis for the conclusion that salt is soluble, and you could use that information to guide how you used salt, even without pure water. The picture we have is that the structure that sustains dissolving in pure water is still there, and it still affects the behavior of a substance even when there is no pure water and the only water we have is muddy or otherwise impure. Similarly, if we think of causation in terms of counterfactuals about what would happen under ideal interventions, it seems as though we're thinking of it as an underlying structure that can also be revealed under imperfect interventions, knowledge of which can be put to good use in acting even outside the context of ideal interventions.

Relatedly, when we find that the results of a well-executed randomized controlled trial show that X causes Y—when they show, for example, that the action of a particular drug prevents heart disease—then we say, "There must be a mechanism." We take it that such trials can establish causation, but we also take it that there is further work to be done to establish the mechanism by which X causes Y. Incidentally, notice that it is not obvious what justifies this further step when we say, "There must be a mechanism." There does not seem to be any contradiction in the idea that it could be that interventions on X are correlated with changes in Y, even though there is nothing describable as a mechanism to be found. Nonetheless, in the physical case(and certainly in the case of, for example, medicines), most scientists and indeed most educated people generally would have a strong a priori view that there must be a mechanism.

How does it go for the case of randomized controlled trials, or ideal interventions generally, involving psychological variables? Again, we assume that we can demonstrate causality on the basis of "imperfect evidence," and we assume that the discovery of a causal relation has implications that go beyond the context of ideal interventions. For example, studies into the buying behavior of consumers, or the causes of voters choosing one candidate over another, are typically based on imperfect evidence and then used in contexts other than those of ideal interventions. We again seem to be working with the picture of an underlying structure that is not revealed only in the context of ideal interventions. And suppose that psychological variable X is demonstrated to cause psychological variable Y. Suppose, for example, that insomnia is found to be a cause of depression. Suppose we have excellent experimental evidence for this connection. Do we again assume that there must be further work to be done to discover the mechanism? In the psychological case, do we again have the strong a priori conviction that "there must be a mechanism" when two variables are found to be causally related? In the next chapter, we'll begin on what the right notions of mechanism and process might be here. But now I want to look at one further reason why an interventionist approach to causation may be thought to require concepts of mechanism and process.

5. THE CONCEPT OF INTERVENTION

Let's go back to the concept of an intervention. As I said, the following definition is canonical in the philosophical literature, building on the earlier work by Spirtes, Glymour, and Scheines (1993) and Pearl (2000).

> I is an intervention variable on X with respect to Y if and only if:
>
> I causes X
> I does not cause Y otherwise than by X
> I not correlated with any Z causally relevant to Y otherwise than via X
> I suspends X from the effects of the factors that usually impact it (Woodward 2003, cf. Hitchcock and Woodward 2003)

The interventionist defines "X causes Y" in terms of there being changes to the value of Y under some (possible) interventions on X. But notice how heavily this definition of intervention itself uses the notion of cause. It's fair to reflect that a definition can be illuminating though circular, and that the definition does not use the idea of a causal connection between X and Y, which is what is being defined (cf. Woodward 2003). But the circularity of defining cause in terms of intervention and intervention in terms of cause leaves us with two problems, one epistemic and one conceptual.

The epistemic problem is: With what right can we ever assume that we've demonstrated causation? To do that, we'd have to show first that there was a successful intervention on our target variable X. But to do that would require establishing the various points about what was causing what listed in the definition above. At this point, we seem to have an infinity of points to establish because to determine what was causing what, we'd have to consider what would happen under a fresh set of interventions, and so on. One natural answer is suggested by the following remark of Clark Glymour's.

> The implicit assumption of freedom of the will is essential to learning. If we did not at least unconsciously assume our own actions to be autonomous, we could not learn the effects of our own actions; and if we did not assume the same of others, we could not learn the effects of our own actions by observing theirs. If, in action taken or observed, the application of that assumption is conscious, we must have the illusion of conscious will. (Glymour 2004, 262)

The idea is that the way we solve the epistemic regress is to implicitly take it that our own actions, and the actions of others, are interventions in the sense of the definition above. They are autonomous in the sense that they're not caused by external factors that are also causes of the outcomes we're observing. The idea that children learn about causation by implicitly taking their own actions to be ideal interventions is pursued in Gopnik et al. (2004). This is quite a persuasive idea. It leaves open the possibility that in any particular case, you could defeat a claim to have demonstrated causation by showing the actions of an experimenter were not ideal interventions. But the default is the implicit assumption that we are in the good case.

Whether you could justify such an implicit assumption, and how, will be a question similar in depth and importance to addressing philosophical skepticism about perception. We do implicitly assume that we are in the good case, and everything depends on that. But it's a further question with what right we do so.

In this chapter, we've seen that at the level of general causation, we can find causal relations among psychological variables across a population. But what we will see in the next chapter is that we can't use the experimental method to find the causal relations between particular psychological events. One way to put the question is to ask whether this agentive picture gives a persuasive model of one's relation to one's own mind. Here you are, in introspection, confronted with the gadgets and switches, wires and pulleys, and levers and light bulbs of your own mental apparatus. Can we think in experimental terms of your perspective on the causal relations between your own mental states? And can we think in experimental terms about your knowledge of the causal relations among someone else's mental states? We'll see that we can't, and that means we still have the epistemic problem—how we can establish singular causal relations among particular mental states.

I said that there were two problems raised by the circularity of defining cause in terms of intervention and intervention in terms of cause. The second problem is the conceptual one. What is it to understand the concept of cause? Once someone gives an explicit definition of a concept, the natural thing to say is that understanding the concept is a matter of knowing the definition. But if the definition is circular, that doesn't work. If people don't understand the concept of cause, you can't explain it to them by providing them with the interventionist definition. The natural answer is again an appeal to agency. You exhibit your understanding of the concept of cause by implicitly treating your own actions, and those of others, as ideal interventions. But again, if the thrust of the discussion so far has been correct, that approach simply won't work for the case of singular mental causation.

If the agentive strategy isn't available, what other responses could we have to the epistemic and conceptual problems? Another direction to go in would be to try to remove the circularity from the definition of "intervention." We might hope to find some notion of a causal process that we could define without appealing to the concept of causation. We could then use that notion of process in defining the concept of intervention. For physical causation, this kind of idea has been explored by many writers, such as Fair (1979) and Dowe (2000). I'm not aware of anyone doing this explicitly for the mentalistic case, but on the face of it, a parallel approach would be possible. In the physical case, we would define "I is an intervention variable on X with respect Y" as follows:

A physical process links I into X.

No physical process linking I into Y excluding X.

I not correlated with any Z linked into Y by a physical process
   excluding X.

I suspends all other physical processes linked into X.

This is obviously only the rough draft of an approach, but we can
already see how it might go in the mentalistic case.

A mental process links I into X.

No mental process linking I into Y excluding X.

I not correlated with any Z linked into Y by a mental process
   excluding X.

I suspends all other mental processes linked into X.

Of course, we would now need to explain the underlying concept of process. On the face of it, it seems unlikely that we'd be
able to find a definition that applied univocally to both the mental and the physical cases, so we'd have found a certain disunity in
the concept of cause.

Nothing I've said so far challenges the idea that we could have
first-grade interventions of exactly the same sort in both the mentalistic and the physical cases. That is, if you want to conduct a
randomized controlled trial of the causal implications of the belief
that someone is a librarian on what else people will believe about
her, you can use exactly the same methodology as you use in con-
ducting a randomized controlled trial to find the effects of aspirin
on heart disease. But this raises a puzzle as to how that formal
sameness of methodology for determining general causal propositions can be reconciled with the apparent differences we shall find
between singular causation in the mental and physical cases.

CHAPTER TWO

SINGULAR CAUSATION

Although the way we discover general truths about psychological causation does not use the idea of rationality, the idea that psychologically explaining an action is a matter of showing how the action was justified continues to have great appeal to philosophers. Here is Davidson again:

> Because justifying and explaining an action so often go hand in hand, we frequently indicate the primary reason for an action by making a claim which, if true, would also verify, vindicate or support the relevant belief or attitude of the agent. "I knew I ought to return it," "The paper said it was going to snow," "You stepped on *my* toes," all, in appropriate reason-giving contexts, perform this familiar dual function. (1980, 8)

In this chapter, we'll look at how the idea of rationality has a role to play in our understanding of singular causation in the mental, as opposed to the level of general causation that we've been concerned with so far.

1. TWO TYPES OF MENTAL CAUSATION

In an influential paper, Ned Hall (2004) argued that with physical causation, we need to distinguish between cases of singular causation in which we have mere counterfactual dependence of the effect on the cause (without the cause, the effect wouldn't have happened), and cases of singular causation in which the cause produced the effect. I argue that we can make a parallel distinction in psychology between cases in which the effect of a mental cause is merely counterfactually dependent on it and cases in which the mental cause produced the effect. It is in understanding the idea of one mental event producing another that we need the idea of a rational progression, or more generally, something of which we can achieve imaginative understanding. As we'll see, the notion of a rational progression or justification of a psychological outcome is merely a special case of the more generic concept of psychological process that we need in explaining mental causation.

1. Omission. Consider the following case. Suppose you once told me that your heaviest work week is in September, but I have forgotten all about that. I'm trying to organize a conference that I hope you'll attend, and I wonder whether September would be a good time. Thinking about it now, I surmise that September might be a good time for you—not too far into term, not too clogged with administration yet—so I schedule the thing for September. My decision to set the time for September was counterfactually dependent on my forgetting that your heaviest work week is in September. If that's our notion of cause, my forgetting is a cause of my deciding to set the conference time for September. Forgetting is a psychological state and is part of the psychological explanation of why I acted as I did. But does my forgetting rationalize my decision? Does it in any way justify my decision? Not at all: that I forgot such an important piece of information actually undermines my decision, rather than underwriting it. The intuitive picture here is that there is a rational process that generates my deciding to set the meeting time for September. But my forgetting is not itself part of that rational process. It stands outside the process. Nonetheless, it was a cause of my decision, in that I would not have made that decision had I not forgotten.

There is a parallel between this case and a family of counterexamples provided by Hall (2004) to the idea that physical causation requires contact. Here the principle "no action at a distance" plays the same role as does the appeal to rationality in the discussion above. Hall pointed out that in the case of an omission, for example, there is no sequence of spatiotemporally continuous processes connecting the omission to the outcome. If I don't press the lever in the signal box that puts up the sign telling the train to stop, the subsequent collision may be counterfactually dependent on my omission. It wouldn't have happened if I'd pressed the lever. In that sense, the omission caused the collision, and ordinarily we'd talk in that way. But there were no spatiotemporal processes connecting the signal box to the collision. So if we say the omission caused the collision, it seems too cheap to say this is a counterexample to "no action at a distance." Hall's reaction, and that of other writers (cf. Sober 1985), has been to say that there are two concepts of causation: causation as production and causation as counterfactual dependence.

To get the point over, I've given the simplest case in which we have counterfactual dependence (of my decision to set the time for September on my forgetting) without production (my decision wasn't generated by the forgetting; the forgetting was part of the frame for the production of my decision by other beliefs and wants of mine). Forgetting seems like an omission in that it's a matter of some relevant information not being available to me in the justification of my action, rather than my having something available to me that helps justify the action. As Austin (1957) made plain, there are many further, subtler types of example illustrating the general distinction between, in Austin's terms, psychological states that justify an outcome and the psychological states that in one way or another excuse it.

2. *Double Prevention.* Suppose that hooligans remove a stop sign from a busy intersection late one night. At noon the next day, there's a car crash at that intersection. If the sign hadn't been removed, the crash wouldn't have happened. Yet there's no local process connecting the removal of the sign to the crash. The situation is rather that the removal of the sign prevented the operation of a process that would have prevented the crash. Did the removal of the stop sign cause the crash? A natural reaction is to say that our ordinary concept of cause covers two distinct phenomena: counterfactual dependence, and the operation of causal processes. (For the example, see Hitchcock 2001; for that reaction, see Hall 2004.)

Are there such cases of double prevention in psychology? Suppose Ting-An is wondering whether to take a post that would require her to spend a lot of time in China. In the hallway, there is a newspaper with a story about how difficult life can be for foreign workers in China, suggesting that they can be subjected to arbitrary court practices and jail time. Had she read that article, Ting-An would likely have refused the job. But someone comes by and takes away the paper before she can get to it. She accepts the job offer.

Was the removal of the newspaper a cause of Ting-An taking the job? Her acceptance of the job was counterfactually dependent on the removal of the paper. Had it not been removed, she would not have taken the job.

However, there is no causal process, in any intuitive sense, connecting the removal of the newspaper to Ting-An's decision. Not a single photon, for example, was deflected from the paper to Ting-An's retina. The removal of the paper did not initiate a process culminating in her decision.

Let's now look at an internalized version of that same case. Suppose, as before, that Ting-An has the offer of a China post, but she hasn't yet had time to reflect on it at all. Quite independently, she is thinking about those disturbing reports one hears about the treatment of foreign workers in China. She mulls them over and comes to a firm conclusion: they are likely exaggerations, magnified by xenophobia, of real incidents, but probably they are of no concern to most visitors. That evening, quite separately, she finally sits down to think over her job offer. She accepts the post.

Here, as in the case of the newspaper, her decision is counterfactually dependent on the earlier incident. Had she earlier concluded that the reports were likely the tip of an iceberg and that foreigners in China are generally in danger, it would have factored into her evening review, and she would not have taken the post. But in fact, her earlier reflections had extinguished the concern, and she had put it out of her mind.

Here again, although there is counterfactual dependence, it is not as if her earlier reflections on these reports generated a process (in any intuitive sense) culminating in her decision to take the post. Rather, her reflections merely extinguished a concern that would otherwise have played a role, but in fact it did not play any role.

We can use neuron diagrams from Hall (2004) to describe the situation. Arrows connecting neurons indicate stimulatory connections; round-ended connections indicate inhibitory connections. Filled nodes indicate that the event happened; an empty node indicates that the event didn't happen (see fig. 2.1). In the case of Ting-An, *a* is Ting-An's receipt of the job offer (leading to her mulling over and deciding whether to accept it), *b* is the report of conditions in China, and *c* is the removal of the newspaper (or Ting-An's verdict that these reports are to be set aside).

These are cases of double prevention. I think it's instructive to compare and contrast them with the classic case of Suzy and Billy. Suzy is a bomber pilot. An enemy pilot comes on course to shoot

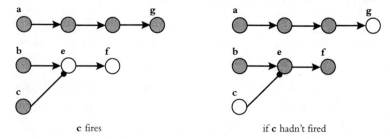

c fires   if **c** hadn't fired

Figure 2.1. Double prevention. Arrows connecting neurons indicate stimulatory connections; round-ended connections indicate inhibitory connections. Filled nodes indicate that the event happened; an empty node indicates that the event didn't happen. Here, "*c* fires" shows a case in which the firing of *c* prevents the firing of *e*, and the causal chain horizontally across the top from *a* to *g* runs to generate the firing of *g*. Then "if *c* hadn't fired" shows that if *c* hadn't fired, the firing of *b* would have caused the firing of *e*, which would have caused the firing of *f*, which would have prevented *g* from firing. Therefore the firing of *c* prevents prevention of the firing of *g*. Ting-An's daytime verdict on reports of xenophobia (*c*) prevents those reports from preventing the job offer (*a*) from generating her acceptance of the offer (*g*). (John Collins Ned Hall, and L. A. Paul, eds., *Causation and Counterfactuals*, figure 4, © 2004 Massachusetts Institute of Technology, by permission of The MIT Press.)

her down. But Billy shoots the enemy down, and Suzy continues to bomb her target. (Here, *a* is Sally flying toward her target, *b* is the enemy pilot getting her on radar, and *c* is Billy firing at the enemy pilot.)

Did Billy's shooting cause the bombing? The bombing was counterfactually dependent on the shooting. There wouldn't have been any bombing if Billy hadn't shot. On the other hand, there may be no sense in which Billy's shooting initiated a process that led to the bombing. The struggle between the enemy pilot and Billy may have happened miles away from Suzy. The structure is the same in the physical as in the psychological case. In both kinds of case, the examples I've given are for illustration only. In both cases, it's easy to multiply examples with the same general structure.

3. Pre-Emption. Let's now look at preemption. In perhaps the single most influential remark in the literature on mental causation, Davidson said,

> A person can have a reason for an action, and perform the action, yet this reason not be the reason why he did it. (Davidson 1980, 9)

The natural interpretation of this remark is that a subject may have two reasons to perform an action and yet perform the action for one reason rather than the other. Perhaps the defendant wanted revenge on the victim and wanted to defend himself against the victim. But the action, when it came, was done out of revenge rather than self-defense, as shown in figure 2.2. On the face of it, this is an example of preemption: revenge preempted defense as a

Figure 2.2. Psychological preemption. You can have both of two motives for an action, revenge and self-defense. If one hadn't generated the action, the other would have. In the case shown, only revenge was in fact operative. But if it hadn't generated the action, the motive of self-defense would have.

Figure 2.3. Physical preemption. There can in in play two ways of shattering a bottle, Suzy's throw and Billy's throw. If one hadn't shattered the bottle, the other would have. In the case shown, only Suzy's throw in fact shattered the bottle. But if it hadn't, Billy's throw would have.

motive. Hall (2004) gave a famous example for the physical case: Suzy and Billy throwing rocks at a bottle. Suzy's rock gets to the bottle first, a split second before Billy's rock whistles through the empty air where the bottle had been. Suzy's rock, unlike Billy's, caused the shattering of the bottle (see fig. 2.3).

Davidson was using his example to illustrate the point that someone having a reason that justifies an action is not enough to show that reason explains the action. That reason further has to be the cause of the action.

Hall was using his example to illustrate the distinction between counterfactual dependence and causal production. On the face of it, the shattering of the bottle was not counterfactually dependent on Suzy's throw. If Suzy hadn't thrown, then Billy's rock would have got the bottle anyway. But the basic point here is that Suzy's rock hit the bottle; there was a continuous process from Suzy's throw to the shattering of the bottle. There was no such continuous process from Billy's throw to the shattering. That's why Suzy's throw is the cause and Billy's is not.

The same comments apply to Davidson's example. We seem to have a distinction between counterfactual dependence and causal production. Perhaps even if the defendant hadn't wanted revenge on the victim, the need for self-defense would have made him strike. Nonetheless, there was a process relating the motive of revenge to the action, and there wasn't such a process linking the motive of self-defense to the action. That's what makes it the case that the action was done for revenge rather than in self-defense.

Hall's argument for the need for a concept of physical "causation as production" was that we need to distinguish between two types of counterexample to the principle "No action at a distance."

1. One type is provided by Newtonian gravity, or quantum nonlocality.
2. The other type of counterexample is provided by causation by omission, or by double prevention.

To deal with the second type of counterexample, we have only to distinguish between causation as production and causation as counterfactual dependence.

Similarly, I've been arguing that in the case of mental causation, we need to distinguish between two types of counterexample to the idea that psychological explanation is a matter of displaying the rationality of the outcome.

1. One type is provided by weakness of will, or the delusions of a psychiatric patient, or cases such as the widower who in grief rolls around in his dead wife's clothes (Hursthouse 1991). These are cases of psychological causation that don't seem to be exercises of rationality.

2. The other type of counterexample is provided by causation by omission, or double prevention.

In the psychological as in the physical case, to deal with the second type of counterexample, we need to distinguish between causation as production and causation as counterfactual dependence.

The other kind of consideration is causal preemption, which in both the physical and mental cases seems to show that causation can't be explained as counterfactual dependence. The case of preemption has been extensively pursued for the case of physical causation; construction of parallel arguments and proposals for the case of mental causation is a fruitful task that I will not pursue here (cf. Schaffer 2016 for a review of examples developing preemption for the physical case).

2. CAN WE ANALYZE SINGULAR PSYCHOLOGICAL CAUSATION IN TERMS OF INTERVENTION COUNTERFACTUALS?

So far in this chapter, we've been looking at the need for a conception of causation as production, as opposed to merely counterfactual dependence, when considering mental causation. Another way of approaching the same point is to consider the kind of analysis we mentioned in the last chapter, of causation as a matter of counterfactual dependence under interventions.

If we do try to explain singular causation in terms of counter-factuals, the natural proposal is that "A caused B" is the same as "had A not happened, B would not have happened." The immediate problem for this proposal is that there might have been a common cause, C, of both A and B. For example, consider the conditional, "If the barometer hadn't fallen, the storm wouldn't have happened." Well, if the barometer hadn't fallen, that would be because the air pressure hadn't dropped, and in that case, the storm wouldn't have happened. The conditional could be true even though the barometer didn't cause the storm—that is, even though A didn't cause B. The natural response is to consider a counterfactual about what would have happened under an intervention on the position of the barometer needle: had you reached in and pushed the needle to a new position, would there have been a difference in whether the storm happened? If not, then the barometer reading isn't a cause of the storm.

This is surely the right way to develop a counterfactual analysis of causation. The idea runs into problems with the examples we have considered, particularly with cases such as preemption. But in this section, we'll see that there are further problems when we consider the use of the idea of an intervention in this context. In the psychological case, what is the analog of reaching in from outside and moving the needle on the barometer?

I. INTERVENTIONS ON OTHERS

Suppose we start by looking at your knowledge of the causal relations among someone else's mental states. We ask how interventionist approach applies to your understanding of what is caused by someone else's particular belief. For instance, we question whether Sally's belief that there is mud in the water caused her to throw it away. Maybe she also believed that the water was no longer needed. Which was the cause of her action? On the interventionist analysis, the question has to do with what would happen under interventions on Sally's belief that there is mud in the water. That belief has to be moved as one would a lever, so that we can find the upshot. If whether she throws the water away is correlated with whether she believes that there is mud in the water, under interventions on that belief, that is what it is for there to be a causal connection between the belief and the action.

Yet what would it be for you to seize control of Sally's belief that there is mud in the water? We can take it that there is in place a distinction between the variables endogenous to the system we are studying—Sally's psychology—and the exogenous variables, such as your actions on Sally. The intervention must come from outside and seize control of whether Sally has the belief. As I said, the intervention has to rule out the possibility that there is a background common cause of Sally's belief and her action that explains a covariation between them. Whether Sally has the belief must be suspended from the influence of its usual causes, such as her background reasoning from other beliefs she has.

This would obviously be an unusual situation. It does not happen very often that one person is able to reach into another's mind and take control of one's belief. Moreover, when we act on a belief, we typically keep the belief under review while executing the action. Suppose I think the big box will be relatively easy to lift, and on that basis, I try to pick it up. If I have difficulty executing the action, I may revise my belief. However, if someone else has reached in and seized control of the belief, I will be unable to do that. In that sense at least, my acting on the belief will no longer be an exercise of rationality on my part. In fact, the belief has so far parted company from my reasons for it that we would naturally say that it is not my belief at all but rather something implanted in me by the intervener.

It is difficult to accept that our interest in whether one belief rather than another is causing a particular outcome is an interest in what would happen in this unusual scenario. When we are concerned with psychological causation, one topic that is of importance is causal connections among propositional attitudes that reflect the autonomous thinking of the subject. But on this third-person interventionist approach to singular mental causation, we seem to lose sight of anything but scenarios in which the autonomy of the subject has been short-circuited.

2. INTERVENTIONS ON ONESELF

You might conclude from this that if we are interested in an interventionist approach to mental causation, we should think in terms of what happens under one's own interventions on one's own psychological life. What does it come to that you grasp the causal role of your own belief? In particular, what does it come to that you grasp the possibility that one of your beliefs may cause you to have further beliefs, to act in particular ways, and so on? On the interventionist account, what it comes to is this: you grasp that interventions on your belief would be correlated with changes in your further beliefs and with changes in your actions. An exogenous cause tweaking your belief would be correlated with changes in your further beliefs and your actions.

This version of interventionism takes your own actions on your own mental states to be the paradigmatic exogenous manipulations. On this approach, the way to understand the significance of a proposition "the belief that this letter is addressed to me caused me to open it" is something like this:

> My own manipulation of my belief, "this letter is addressed to me," is correlated with changes in whether I open it.

For that to work, you would have to be capable of manipulation of your own beliefs. You have to reach into the system of mental wires and pulleys and levers. You would have to be able to manipulate your own belief to see what happened next. But that is not possible. You can't decide what to believe. There is, as Williams (1970) once pointed out, arguably a principled impossibility about this. Belief intrinsically aims at the truth. For that reason, beliefs can't simply be manipulated at the will of the agent. Any state that could simply be manipulated at the will of the agent would, for that reason, not be a belief. But this means that we can't explain an understanding of the causal significance of one's own beliefs in terms of an agentive version of interventionism.

Of course, there is some discussion about the possibility of deliberation as to what to believe (cf., Shah and Velleman 2005). Whatever we say about the extent of deliberation as to what to believe, this deliberation cannot amount to an intervention on one's own belief. An intervention on one's own belief would have to suspend the usual causes of the belief from affecting it, but of course what one does in deliberation is to weigh the usual reasons for belief, not to somehow set them aside.

You can manipulate the distal causes of your belief (the world itself about which you have beliefs), but that isn't the same thing as you intervening on the belief itself. Manipulating the outside world itself might well result in changes in your beliefs. If you change whether there's mud in the water, that will typically change your beliefs on that point. But what the interventionist analysis of causation requires us to consider is a manipulation that directly affects the target variable itself (see fig. 1.1). Manipulating an aspect of the world itself leaves open the possibility that the aspect of the world you are manipulating might be a common cause of both the belief itself and the outcome variable. We need to know what in principle would happen to the outcome variable under manipulations of the target variable itself. The role of the agent in manipulating the outside world rather than the target variable (belief) directly is not relevant to question what constitutes a causal connection between the target variable (belief) and the outcome variable (further beliefs or action). Again, interventionism does not explain what grasp of the concept of cause comes to in this case.

3. PERCEPTION AS A NATURAL EXPERIMENT ON BELIEF

Interventions need not always be exercises of agency. There can be natural experiments, in which something happens to A not because of anyone's intentional action, and we can see what that makes happen to B. But in the case of mental causation, the kinds of interventions that seem central are not necessarily natural experiments on beliefs. I think a better model is provided by another kind of experiment—namely, that in which we try to set up a situation so that some external magnitude intervenes on the state of our measuring instruments. Ordinarily, such experiments don't constitute interventions on beliefs because we keep in play our capacity to challenge the correctness of our measuring instruments. But there seems to be something fundamental about the role of perception itself as making possible exogenous impacts in which the world intervenes on one's belief. Agency may be involved because often enough in perception, you are looking to find the answer to a question. These aren't always natural experiments. But agency comes in as you setting yourself up to be intervened upon rather than as you setting yourself up to intervene on something else.

Suppose you are sitting in a chemistry class. You are watching an experiment and are going to write a report. You have it on the authority of the textbook and the teacher that the liquid in the test tube will turn yellow. You believe that the contents of the test tube will turn yellow. When you write down your report, what is the cause of your action, reporting the liquid to be one color rather than another? Are you a mere mouthpiece of the official view, writing down whatever color the text and the teacher specified? Or did you write down the color you did because that was the color you believed the liquid to be? So long as the values of all three variables are correlated, there is no way of applying the distinction (see fig. 2.4). You might claim to have some insight into your own motivation here. You might protest that you are not a mouthpiece. What you write down causally depends on what you believe. Text and teacher enter into the proceedings only because you believe them. You are not merely writing down whatever they say, whether or not you believe it. Well, what is the difference between these two hypotheses? The difference shows up when there is an intervention on belief. Some external factor has to take control of your belief, suspending it from the influence of these other factors. In this situation, we can look at whether your belief and action are correlated. The holding of the causal relation consists of them being correlated under such an intervention.

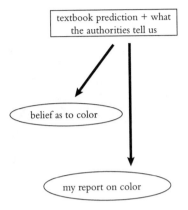

Figure 2.4. Causation or correlation? My beliefs as to the color of the liquid in the test tube are correlated with my reports of its color. But that need not be because my beliefs are causing me to write one way or another. Rather, they may both be effects of a common cause.

In perception, ordinarily, the world itself intervenes on your belief, suspending it from the influence of those prior expectations. The critical point comes when the test tube itself is displayed, and you see that it is, for example, bright blue. What then do you write down? Does your action vary with the belief? Or does it vary rather only with the specification of color by text and teacher? If, under the intervention on belief provided by the world in the case of perception, your report is correlated with what you believe, then the belief is causing the report (see fig. 2.5). You might suggest that our ordinary understanding of the causal relations between our own mental states depends on grasp of this role for perception as providing for interventions on belief. Our grasp of the role of perception shows up in what we would ordinarily say about a classroom full of people who wrote down "yellow" in the situation I have described. We would say that they are not writing this down because they believe the contents of the test tube

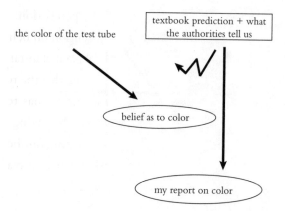

Figure 2.5. Perception as an intervention on belief. In ordinary perception, the world itself intervenes on your belief as to how things are. We can determine whether your report is caused by your belief by having the world intervene on your belief and seeing whether, in that condition, differences in the world correlate with differences in your report.

are yellow. We would say they are writing this down because that is what the officials say they should write down. If I recall that I myself wrote down "yellow," I would have to acknowledge that this was not because of my belief about the contents of the test tube. (For discussion of the implications of the status of perception as an intervention for belief updating in a Bayesian framework, see Glymour and Danks 2007.)

It does not seem likely that this point will give us enough traction to constitute a fully general approach to mental causation. After all, there are many beliefs that can't be intervened on by perceptions. In the case of simple perceptual beliefs, it seems arguable that the belief is genuinely intervened on by the perception; it may have no other cause than the perception. But most of our beliefs seem to be influenced by perception rather than being seized by it, and it's very hard to see how you could begin to think of interventions on desires as being driven by perception. So far as I can see, the role of perception as an intervention on belief is quite special and not something of which we find parallel instances across many different types of mental state. It may be that the role of perception as an intervention on belief specifically has to do with our human need for a reality check on our theorizing: that perception has to play a decisive role in determining our beliefs if we are not simply to be overwhelmed by confirmation bias and our capacity to generate endless further theorizing.

3. IMAGINATIVE UNDERSTANDING OF PSYCHOLOGICAL CAUSAL PROCESSES

It is when we consider our knowledge of the psychological processes implicated in singular causation that we see the appeal of an approach to the dynamics of the mind that emphasizes the role of imaginative understanding. It has often been recognized that imagination plays a role in providing us with knowledge of what's possible, of what could happen. But what we have to consider here is the role of imagination is providing us with knowledge of what is actually going on. The role of imagination in providing knowledge of the mind has sometimes been recognized in analytic philosophy. But what will matter for us is what we might call the dynamic role of the imagination, the role it plays in an understanding of how one mental state generates or produces another. In contrast, there is simply getting a snapshot—a static picture—of how someone's subjective life is qualitatively at some one moment in time. This is the usual interpretation of Nagel's famous example of the bat: that it's pointing up the way in which our imaginative understanding of one another usually gives us this kind of snapshot knowledge of each other's experiential states, and that is so strikingly missing in the case of the bat (Nagel 1974). Focusing on this static aspect of the imagination misses an at least equally basic role: your knowledge of the dynamics of someone else's mind providing you with knowledge of how one particular psychological state generates another, as when you see how your friend's grief has led to anger. Nagel's example is the simplest sharp statement of the need for imaginative understanding. Suppose you're trapped in a barn with an excited bat, and you raise the question what its subjective experience is like. Nagel's point was that no amount of scientific investigation will provide you with that information. You can pull the bat apart cell by cell, for example, and develop a complete computational account of how it's responding to the physical structures and forces around it, but you still have no idea what its experience is like. Nagel put his main point like this:

> At present we are completely unequipped to think about the subjective character of experience without relying on imagination—without taking up the point of view of the experiential subject. (Nagel 1974, 449)

Insofar as we don't have an imaginative understanding of the bat, we don't have any knowledge of its qualitative states. It's only imaginative understanding that provides our knowledge of what is actually going on with one another's subjective experiences. Ordinarily we do know about one another's subjective lives, and from a practical point of view, that is certainly the most important knowledge we have—does anything else matter at all, except insofar as it bears on this? This knowledge cannot be supplied by science in the case of humans any more than in the case of bats. It is supplied by imaginative understanding.

The dynamic understanding of singular causation in the mind that's provided by imagination is somewhat different. It's knowledge of the psychological processes that underpin counterfactuals we might accept about someone's mental states. The great theorist of this dynamical role for the imagination is Karl Jaspers. Jaspers was primarily concerned with psychiatry and what goes on in the clinical encounter between therapist and patient. But the distinction he draws between two different ways in which the therapist can engage with the psychology of the patient is applicable quite generally and is central to our current question, which is how to think of our conception of psychological causation in terms of processes rather than merely counterfactual dependence.

Jaspers draws a distinction between two approaches to the study of patients, (1) "subjective psychopathology (phenomenology)" and (2) "objective psychopathology."

1. We sink ourselves into the psychic situation and *understand genetically by* empathy how one psychic event emerges from another.
2. We find by repeated experience that a number of phenomena are regularly linked together, and on this basis *we explain causally.* (Jaspers 1913 / 1959, 301)

Now we can think of the objective psychopathology that Jaspers describes under (2) as a brief description of the kind of recovery of causation from knowledge of statistical regularities, particularly regularities about what happens under interventions, which we reviewed in Chapter 1. What matters is what happens to Y in a population when there's an intervention on X in that population. But what about Jaspers's first type of approach to psychopathology, using empathy? Empathy provides us with grasp of "meaningful psychic connections." These connections, Jaspers says, "have also been called 'internal causality,' indicating the unbridgeable gulf between genuine connections of external causality and psychic connections which can only be called causal by analogy" (1913 / 1959, 301). Let's look further at our knowledge of this internal causality between psychological states.

> In the natural sciences we find causal connections *only* but in psychology our bent for knowledge is satisfied with the comprehension of a quite different sort of connection. Psychic events "emerge" out of each other in a way which we understand. Attacked people become angry and spring to the defense, cheated persons grow suspicious. The way in which such an emergence takes place is understood by us, *our understanding is genetic.* Thus we understand psychic reactions to experience, we understand the development of passion, the growth of an error, the content of delusion and dream; we understand the effects of suggestion, an abnormal personality in its own context or the inner necessities of someone's life. Finally, we understand how the patient sees himself and how this mode of self-understanding becomes a factor in his psychic development. (Jaspers 1913 / 1959, 302–303)

So does imaginative understanding constitute knowledge of causal relations? On the one hand, it's not knowledge of external causal relations and is called "causal" only by analogy. On the other hand, it is genetic and constitutes knowledge of how one psychological state emerges out of another. I think that Jaspers's point has to do with how we have a conception of psychological causation that is not merely a matter of correlations, not merely a matter of probabilities or counterfactuals about what happens under interventions. We are able to think in terms of causal processes operating in the mind. Let's look again at the analogy with the physical case. To vary a traditional example, suppose that you're watching as Moriarty, standing at the top of the Reichenbach Falls, is about to push a rock over the edge of the cliff, and Holmes is standing at the bottom (cf. Good 1961–1962; Hitchcock 1995). Moriarty is a deadly rock-pusher, and if he gets to aim, Holmes is done for. At the last minute, Watson rushes up. It is too late for him to seize the boulder from Moriarty; all he can do is give it a wild shove. The boulder bounces erratically from side to side before eventually crushing Holmes. As an observer watching this, you know definitively what caused Holmes's death: it was Watson's pushing the boulder. No doubt there are laws of nature without which there wouldn't have been this cause and effect sequence, but they, and the initial conditions of their application, may be below your radar as an observer. The sequence may be chaotic, in that differences in the initial conditions too small to measure—a slight change in the direction of Watson's push, a difference in the position of a blade of grass—may make large differences to the outcome. This gives us an analogy from the physical case for what Jaspers calls understanding a genetic connection in a particular person's mind. For Jaspers, the analog of following the trajectory of the boulder is following the meaningful generation of one psychological state from another. Jaspers writes:

> The self-evidence of a meaningful connection does not prove that in a particular case that connection is really there or even that it occurs in reality at all. (Jaspers 1913 / 1959, 303)

What is required in imaginative understanding is that you should grasp how one psychological element emerges out of another in a way that is meaningful. But merely grasping a meaningful relation between two elements of another person's psychological life is not enough to establish that one was the genesis of the other.

> Nietzsche convincingly and comprehensibly connected weakness and morality and applied this to the particular event of the origin of Christianity, but the particular application could be wrong in spite of the correctness of the general (ideally typical) understanding of that connection. In any given case the judgment of whether a meaningful connection is real does not rest on its self-evident character alone. It depends primarily *on the tangible facts* (that is, on the verbal contents, cultural factors, people's acts, ways of life, and expressive gestures) in terms of which the connection is understood, and which provide the objective data. All such objective data, however, are always incomplete and our understanding of any particular, real event has to remain more or less an interpretation which only in a few cases reaches any relatively high degree of complete and convincing objectivity. (Jaspers 1913 / 1959, 303)

However provisional our findings have to be in the psychological case, in principle we can follow how one psychological state generates another. Just as in the case of the boulder being shoved off the cliff at the Reichenbach Falls, our ability to grasp the reality of the imaginative connection—our ability to find how one psychological state generated another—does not depend on knowledge of regularities.

> Because we note the *frequency* of a meaningful connection
> this does not mean that the meaningful connection becomes
> a rule. This would be a real mistake. Frequency in no way
> enlarges the evidence for the connection. Induction only es-
> tablishes the frequency, not the reality of the connection it-
> self. For example, the frequency of the connection between
> the high price of food and theft is both understandable and
> established statistically. But the frequency of the understand-
> able connection between autumn and suicide is not confirmed
> by the suicide curve, which shows a peak in the spring. This
> does not mean that the understandable connection is wrong since
> one actual case can furnish us with the occasion to establish
> such a connection. (Jaspers 1913 / 1959, 304)

The general point here is that you can have knowledge of singular causation in psychology that doesn't depend on there being (in
any practical way) replicability, and that only imaginative understanding seems to provide this. There's an analogous point to be
made about knowledge of singular causation in the physical case.
When you're watching Watson push the boulder to crush Holmes,
you may know definitively what caused the death of the individual crushed even though there may be no practical possibility of
replication. And it's not as though replication would help, anyway. "Frequency in no way enlarges the evidence for the connection." You already know what killed Holmes. Seeing someone
push another boulder, in a neighboring valley, and cause another
death would simply provide some knowledge as to how often this
kind of thing happens, not bolster for the existence of the causal
connection in the original case. You can watch and imaginatively

understand the autumn bringing about winter blues, and ultimately depression and suicide, in someone you know well. Your
imaginative understanding here does not depend on your knowledge of some generalization about autumn and suicide, and in
fact there may be no such generalization to be had.

This way of looking at things echoes a traditional distinction
between nomothetic explanation and idiopathic understanding.
Perhaps we can make sense of imagining what it would have been
like in seventeenth-century London, or imagining life for a Sunni
Muslim in Iraq today. In this kind of case, though, what one
imagines (the kinds of connections you imagine there being in
the psychological life) presumably has to be a schematic version of
what is actually going on in a real person's life. In contrast, in the
external case, the relations of general causation between variables
need not be instantiated as relations of singular causation.

Jaspers said, "The most profound distinction in psychic life
seems to be that between what is meaningful and *allows empathy*
and what in its particular way is *un-understandable,* 'mad' in the
literal sense, schizophrenic psychic life" (1913 / 1959, 577). Jaspers famously thought that the delusions of a psychiatric patient
are "un-understandable," so their genesis can't be grasped as meaningful. I'm suggesting the notion of meaningful connection is
what we need to explain the parallel between our knowledge of
singular causation in the physical case and our knowledge of singular causation in the mental case. In the physical case, our notion of physical process seems to depend on our grasp of some
such ideas as sameness of physical object and spatiotemporal con-
tiguity. We have seen that the analogous concept we need for the
mental life is that of a meaningful sequence of psychological
events, though this isn't quite the notion of "meaning" that the

term may suggest to a philosopher. As Jaspers uses the term, it seems to be correlative with the possibility of imaginative understanding, following the genesis of one psychological event from another.

There is one qualification I want to make here. Suppose you consider watching one of the early Heider and Simmel sequences, in which a large square chases around a smaller square and a circle (Heider and Simmel 1944). To most of us, it seems compelling to interpret the scene in terms of singular causal relations between psychological states. The aggression of the large square, for example, instantly causes anxiety on the part of the smaller square. Here, it seems that we're dealing with observation of singular mental causation as a simple perceptual phenomenon, like the Michottean launching phenomena (Michotte 1963). That kind of perception of singular causation may also be at work in some of the cases Jaspers calls "un-understandable." I was talking once to psychiatrist Ken Kendler about Jaspers's talk of "un-understandability," and I said, Well, no matter how crazy or elaborated the delusion, it always seems possible to "understand" it, in a fairy tale or just-so kind of way. The FBI are putting cameras in the patient's shoes and so on, this is why he has to buy a new pair every day.

No, Ken said, that misses what is so often jarringly incomprehensible about delusions. Take for example a patient he knew well with whom he had a disagreement about the correct course of treatment. In the middle of the discussion, she suddenly exclaimed, "You're not Dr. Kendler! Dr. Kendler would never have treated me like this." What is un-understandable here is not what the patient is saying. Neither is it unknown what caused the patient to say that. It was the disagreement between them that was the cause. In fact, the disagreement causing the delusion seems like one of those perceptual "launching" phenomena. What is starkly un-understandable is how the disagreement could have given rise to the delusion. That is what we cannot understand—"genetically by empathy" in Jaspers's term: how it could have been intelligible for the disagreement to give rise to the delusion. In our ordinary knowledge of one another's mental lives, where things are understandable, we follow the trajectories of singular causation through rational thought, just as in physical understanding, we follow the spatiotemporally continuous trajectories of the objects around us.

Indeed, once we have grasped this dynamical role for the imagination in our understanding of the mental life, it seems possible that it's the dynamical conception of imagination that we should have been using, even in thinking about Nagel's bat. One thing we don't know about, in the case of the bat, is to which aspect of the external environment it's responding. We don't know about the things and properties that matter to the bat any more than it has perceptual knowledge of the colors and shapes and people that matter to us. We don't know which aspects of the environment it's experiencing. The other thing we don't know about is what impact particular perceptions have on the rest of the psychological life of the bat. That is the sense in which we'd ordinarily use the phrase "knowing what it's like": if someone says, "You don't know what it's like to be told that you've been fired," the point is not about not knowing what being fired is—some qualitative dimension of the thing—but rather about not having any imaginative understanding of the psychological consequences of having been fired. Or you might say, "You don't know what it's like to see the house you've always lived in burning to the ground." There are two parts to this: knowing about the perceptible characteristics of a burning house, and knowing what the psychological impact of perception of them would be. It seems possible to interpret Nagel's example in terms of the need for a dynamical understanding of causal processes in the mental, as well as the apparent impossibility of achieving that knowledge through science.

Nagel's example usually seems to be interpreted in terms of the impossibility of a kind of static picturing of the bat's inner states, as if the key kind of imagining here was a kind of mirroring or echoing of the bat's internal qualitative characteristics—the bat's "qualia," as people say. This leads to puzzlement as to how to think of these qualia, which seem to be elusive, not themselves being directly observable but not mere theoretical constructs either. But we could dispense with the qualia. I'm suggesting that the way out of this puzzlement is to let go of this static conception of imagining. We should think of what we're missing, in the case of the bat, as a combination of (a) knowledge of the perceptible characteristics of its environment to which the bat's responding and (b) dynamical imagining of what it does to the bat to observe those characteristics.

4. EXPLAINING MENTAL PROCESSES IN TERMS OF PHYSICAL PROCESSES

In everyday life and in the law courts, we give a good deal of significance to questions about mental causation. In Shakespeare's *Othello*, there's a bit where Othello has to explain how a nice girl like Desdemona threw in her lot with a rough type like him. The court feels that only sorcery or drugs can have caused this to happen, and the general feeling is that if that's so, executing Othello would be about right as a response. Othello has to convince the court it wasn't that way and describes how he used to tell Desdemona about his troubled life, from being sold into slavery to his adventures among the Anthropophagi and men whose heads do grow beneath their shoulders. He says that this awoke Desdemona's pity, which in turn gave rise to love: "She loved me for the dangers I had pass'd, And I loved her that she did pity them" (Shakespeare 1603 / 2016, 1.3.149–170). This grabs the imagination of the court: they see how it would have been from Desdemona's perspective, and that is enough to get Othello more or less off the hook. This example illustrates a number of points about our imaginative understanding of one another, our ability to get into one another's heads to see things from the perspective of another:

1. Empathy or imagination is of significant practical importance: it's only through their imaginative understanding of Desdemona that the court acquits Othello. This seems to be a general point about social life: that a lot hangs practically on imaginative understanding. You can lose your friendship with someone because you saw exactly what the brute was thinking when you lost your job, for example.

2. Empathy or imagination can yield the highest level of certainty about someone else's feelings or motives "beyond all reasonable doubt," as they say in law. Such a lot turns, in our own personal lives and in the law, on questions about people's motives and feelings. We assume that our ordinary imaginative understanding of them yields definitive knowledge good enough for us to bet our lives on it being

right, sentencing people to death, or falling in love on the basis of our imaginative understanding of one another.

3. Understanding by getting inside someone's head, imagining the world from another's point of view, seems to be a quite different enterprise from understanding that person scientifically. I suppose Othello could have taken a different tack and approached Desdemona's case from the standpoint of a social scientist who knows about the factors that generally cause a healthy love in people of her type, demonstrating that they were present in her case. But that isn't what he does, and the vivid evocation he gives of her state of mind, with its detail and particularity, seems antithetical to a scientific approach. Othello is not trying to exhibit Desdemona as merely an instance of some general causal facts.

4. Even though it's not scientific, Othello's point has to do with causation: he has to establish that the cause of Desdemona's love was pity rather than sorcery or drugs. He's establishing a point about causation through the evocation of imaginative understanding, with enough practical certainty to determine a matter of life or death, rather than through the use of science. We tend to think of science as the authority of choice over questions of cause and effect, but this case shows that this isn't always so, and in fact that it isn't so in a kind of case that matters to us all, every day of our lives.

There are many more philosophically interesting points to be drawn from this example—for instance, in the larger context of the play, there's a point to be made about the limitations of our imaginative understanding of one another; Othello's imaginative understanding of Desdemona is in fact somewhat more restricted than he realizes—but I think that is enough morals to be going on with. The general point is that our imaginative understanding of one another seems to generate some kind of definitive knowledge of causation that really matters. For example, suppose we have the case in which the defendant in a murder case is acknowledged to have two motivations for the action. He was under attack from the victim, but the victim was also someone who had caused him harm in the past. The defendant had two motives: defense and revenge. Did he act from one motive rather than the other? The courts provide a rich set of examples in which we take it to be possible to know beyond reasonable doubt which psychological state gave rise to which, on the basis of our imaginative understanding of the defendant. In *Causation in the Law,* Hart and Honore are very explicit that the correct procedure here is not some attempt to apply generalizations to the case.

> The statement that one person did something because, for example, another threatened him, carries no implication . . . that if the circumstances were repeated the same action would follow. . . .
>
> [Consider] the threatened person's own statement that he acted because of those threats. It would be absurd to call upon him to show that there really was this connexion between the threats and his action, by showing generally he or other persons complied when threats were made. . . .
>
> This is recognized in the law as well as in ordinary life. If wanted to make sure, in giving evidence, that his reasons for acting were as he claims, an honest witness will not be

expected to produce generalizations, but to attempt to reconstruct the deliberative situation or his "state of mind" at the time. (Hart and Honoré 1985, 52)

It's your imaginative reconstruction of the state of mind (the deliberative situation) of the accused that provides you with the knowledge of causation that governs the life or death of the individual. We give a lot of practical import to that question. The difference in a particular case may be fifteen years of jail time, or life versus death. Or even in a more mundane case, you and I are in a crowded elevator. You stand on my toe. I know you're annoyed with me anyway, but you also like to keep your balance. Which motive caused you to act? My resolution of that question may forever affect my attitude toward you.

The practical significance that we give to our imaginative knowledge of mental processes cannot be understood if we immediately explain mental processes as neural processes. Distinctions among neural processes, considered simply as such, do not seem to have any practical significance at all. Some analogies here may be helpful. We give a lot of practical weight to the question whether another person is in pain. The distinction between someone being in pain and someone not being in pain matters as much as anything else. It seems possible, however, that if we look at the brain states that realize pain and those that don't realize pain, there is no physically remarkable difference between them. Perhaps if we view the brain entirely from the standpoint of physiology, there is no big difference between the two kinds of physical state—a matter of tiny differences in the frequency of firing of a range of cells, for example. Certainly, however it goes for humans, it seems possible that there should be creatures like that.

Lee (2019) has argued that if things do turn out like that, then we would have to abandon our ordinary belief that it matters a lot whether or not someone is in pain. Because the psychological is grounded in the physical, Lee argues, we must be able to discern the distinctions that matter, as distinctions that matter, at the physical level. This does not seem like a convincing argument; at least, most people are unlikely to be convinced by it. It doesn't really matter how tiny, from the point of view of physics or neurophysiology, the difference is between two neural states. If one grounds the subjective experience and the other does not, then that difference is of great practical significance. Compare the idea that tensed facts—such as whether an event is past, present, or future—are grounded in the tenseless facts about the time-order of events, which come before which, and so on. Suppose you have a disagreeable examination coming up. You fall asleep, wake up, and for a moment think, "Was that exam yesterday, or is it coming up later today?" This seems quite important: if the exam is over, you'll feel relief; if it's still coming up, you'll feel dread and anxiety. Parfit has argued, however, that this is irrational: the tenseless time-order of events is the same whether the exam was yesterday or is today, the only thing you might be unsure about is whether the exam is earlier or later than "this utterance," and who cares about that (Parfit 1984)? But in practice, no such line of thought is going to stop people from caring about whether a disagreeable event is over or yet to come. You can argue that the tensed facts are grounded in the objective tenseless facts and that therefore we ought not to care about matters whose practical significance can't be recognized at the objective level, but that simply gives us an unbelievable picture of what we ought to care about. Similarly, Lee's line of argument is not going to stop people caring about whether or not something hurts. I think the right moral to draw is that even if you think you have found that facts of type A are grounded in facts of type B, you can't conclude that the practical or moral significance of type A distinctions is grounded in the practical or moral significance of type B distinctions. It is similar in the case of mental causation. In everyday life and in the law, it matters very much which mental state caused which. As the case of Othello makes evident, we care a lot about which mental state gave rise to which. But the significance that we attach to questions about mental processes can't be explained by trying to ground the significance in terms of distinctions among kinds of neural processes. It seems to matter very much whether it was pity or fear that caused Desdemona's behavior. But if we say this is, in the end, merely a distinction between two kinds of neural process, we lose our grip on why this kind of technical distinction should matter. The brain is very densely interconnected; maybe at the neural level, the distinction between the case in which Desdemona acts out of pity and the case in which she acts out of fear is merely a distinction between one level of electrical activity and another, of no evident physiological significance. It's only at the level of our imaginative understanding of mental processes that we can appreciate why it matters whether she acted out of pity or out of fear.

There are many philosophers and scientists who will argue that the only conception of causal process that we have or need is the concept of a physical causal process. There's only one kind of causal process, and that's a physical process. Mental processes are correctly identified as causal only insofar as they're grounded in physical processes. Though it does afford some theoretical simplification, there are two related problems with this idea.

First, it seems to lead straight to epiphenomenalism. If we think that all causation is a matter of physical process, then we have to accept that mental characteristics are idle in the functioning of those processes. It is possible to argue over the definition of "idle" and to find some form of words that will allow us to say that the mental characteristics are causally significant. But however exactly we are thinking of physical processes, the fact will be that physical characteristics are playing a role in the generation of outcomes that is quite unlike any role played by mental characteristics. If, for example, we think of physical processes as the communication of motion by impulse, then characteristics such as mass, velocity, and charge will play roles quite unlike any roles that could be played by mental characteristics in the generation of outcomes, and it obviously won't help much to shift to thinking of physical processes in terms of the transfer of energy or exchange of conserved quantities or anything else.

Second, as we've seen, we attach enormous practical and moral weight to distinctions between different types of mental causation. Consider a homicide trial: the difference between acting out of hatred and acting out of self-defense can be life or death. But if that is at bottom merely a difference between one set of neural excitations and another, it is hard to see why the difference should carry that kind of weight. Or to take another example, people sometimes argue that drug addiction is a brain disease and should therefore not carry criminal penalties, whereas others argue that drug abuse is voluntary—after all, people do manage to give up—and therefore should be penalized. Or perhaps the matter is more complex than that contrast suggests, and there should be a more complex approach to criminalization of substance use. But we generally treat these as matters of great practical concern. This concern seems absolutely baffling if we regard it as, at its core, a technical matter about how exactly various neural states are physically generating other neural states.

The view I am rejecting here certainly has an appeal: the idea is that there is a certain physicalism implicit in the ways we ordinarily think about the causes of actions. On this view, when a reason causes an action, there must be a chain of spatiotemporally connected events connecting the reason to the action.

An analogy might be provided by the idea that smoking causes cancer. Here's one interpretation of the causation here. Smoking is not itself a biological variable. It is a sociocultural phenomenon, connected to advertising, Gauloises, Marlboro, and so on. If you want to intervene on something to reduce levels of cancer, the natural thing to intervene on is the high-level phenomenon of smoking. Yet there's certainly a biological process by which it produces cancer.

This picture of a physicalism implicit in our ordinary talk of singular causation in the mental is not without precedent. Martin and Deutscher (1966) argued that our conception of memory of a past perceived event is a causal notion; to remember the past perceived event, one has to be causally connected to one's past perception of it. But the kind of causal connection in question here cannot be explained in counterfactual terms; Martin and Deutscher consider a variety of informants, recorders, and implanted recorders to make the point. The only way of explaining what kind of causal connection is required by memory is to appeal to the idea of the memory trace: the impression that the original perception makes on the subject and that is jogged into action when one remembers. Their suggestion was that this conception of the physical process connecting the remembered event to our subsequent memory of it is already implicit in our ordinary conception of memory.

Even if you are not a reductive physicalist, at first it seems like a coherent package to suppose that we can have high-level mentalistic characteristics such as motivations connected to high-level outcomes such as intentional actions by means of biological processes. Will this give us the distinction we need between counterfactual dependence and causal production for the mentalistic case?

The basic problem here is that the brain is a highly interconnected system, and practically any neural event may be presumed to have some kind of neural connection—perhaps not quite to everything else but certainly to very many other neural states. How could we, at the purely neural level, single out which kinds of biological process subserve a genuine mental process, and which are merely neural connections? In the philosophical literature, the topic has been addressed in the context of discussion of deviant causal chains, as in Davidson's example of a climber who intends to let his companion fall from the rope and is so unnerved by having the intention that he lets his companion fall. It may look as though the action was generated by a mental process, but that's not what happened. Nonetheless, there is presumably an underlying neural chain from the motivation through to the letting drop. Peacocke (1979) has given the most developed attempt to explain in purely biological terms this distinction between nondeviant and deviant chains. His account is entirely in terms of the laws governing the underlying biological phenomena. He defines two key notions:

x's being ø differentially explains y's being ψ iff x's being ø is a non-redundant part of the explanation of y's being ψ, and according to the principles of explanation (laws) invoked in this explanation, there are functions . . . specified in these laws such that y's being ψ is fixed by these functions from x's being ø. (66)

We have stepwise recoverability of p from q iff in the explanatory chain from p to q, at each stage, given just the initial conditions of that stage other than the explanandum of the previous stage, and given also the explanandum of the present stage and its covering law, one can recover the explanandum of the previous stage. (80)

This account is trying to echo, at the biological level, the intuitive demand one would make on a psychological process. The trouble is that as Bishop (1981) remarked, there seems to be no reason why we shouldn't find that these conditions on the underlying biology are met even in cases of deviant causation, where we don't have a mental process generating an outcome. It could be, after all, that if we look at a case of deviant causation, such as Davidson's earlier, we find at the neural level

a. the realization of the intention (to drop the rope) is the only neural state that could generate

b. the realization of the climber's nervousness,

and that this neural state is the only one that could generate

c. the climber's dropping the rope.

In this case, Peacocke's conditions of differential explanation and stepwise recoverability will be met, but we still have a case of deviant causation. The general moral here is that the architecture of the underlying biology may not reflect the structure of mental processes in any concisely capturable way.

In fact, there is a more fundamental point here. Suppose you firmly believe that all the facts about mental causation must be grounded in the facts about physical causation. You might then say that even if Peacocke's account is not quite right, something like that must be right, because only if we can give some such account could mental causation be grounded in the physical.

For the moment, I don't want to resist that kind of idea. My point is only that the high-level conception of mental causation that we are using already needs some kind of explanation. We need to be able to say what mental causation is before we can so much as raise the question how it's grounded in the physical. The proposal I've been considering is that we have to appeal to the physical level, even to say what the high-level phenomenon is. My present point is that at the moment, we don't have any way of explaining how that appeal should go.

There is a more general reason for dissatisfaction with an account of mental processes that appeals only to considerations of systematicity and law, which emerges when we think about how we might expand such an account as Peacocke's away from a focus purely on deviant causation in the cases of action and perception to consider thinking processes generally. On the face of it, such a biological account can't explain why rationality has anything to do with causation as production. Perhaps there is some intuitive sense in which rationality is a systematic phenomenon. But there is no evident reason why systematicity at the neural level could reflect only rational connections at the psychological level. We need a conception of psychological causal process; that is the thing that is the target of our imaginative understanding, and to which we attach such enormous practical importance.

Another way to put the point here is to remark that the idea that the only kind of causal process to connect motive to action is a physical process brings us into the range of the classical mind-body problem. Here are some of the arguments that are usually used to bring out the difficulty of supposing, for example, that the sensation of pain is a brain state.

1. You can imagine have the pain without having the brain state; you can imagine having the brain state without the pain.
2. The pain has a kind of simplicity and unity that doesn't seem to be reflected by the complexity of the underlying physical process.
3. You could know all about the underlying physical processes without ever having known about pain. When you do encounter pain for the first time, you acquire new information.

These arguments, and related ones, have been extensively discussed (cf. Byrne 2006). Many philosophers may respond to them by saying that nonetheless, pain is a brain state. What else could it be? I don't want to enter this dispute here. I simply want to remark that similar considerations apply to our imaginative understanding of mental causation.

a. You can imagine one state giving rise to another in the absence of any underlying physical process, and you can imagine the physical process being there without the one psychological state giving rise to another.

b. Imaginatively understood mental causation has a kind of simplicity and unity that seems to be missing in the underlying physical process.

c. You can know all about the underlying physical processes without having realized that one mental state can give rise to another. When you do gain the capacity to imagine one mental state giving rise to another, you acquire new information.

Following the trajectory of someone's psychology, understanding just how one state gives rise to another is commonplace. We're focusing on the thing we find out about in this way: singular causation in the mind. This is the phenomenon to which we attach practical significance. I'm not arguing that in the end, this phenomenon can be given no physicalist reduction; I'm simply stating that there is a mentalistic phenomenon of causation to be reduced.

5. RELATION TO CLASSICAL SOLUTIONS TO THE PROBLEM OF OTHER MINDS

In the cognitive science literature, the capacity for imaginative or empathetic understanding of another person is usually thought of in terms of the ability to simulate another person. This is a matter of taking on board the other person's beliefs and objectives (where they differ from one's own) and seeing what one would do next oneself, if one had those beliefs and objectives. This whole exercise has to be decoupled from action because otherwise, one would actually be executing actions and behaviors that are in the service of the other person's motivations rather than one's own. What you have to do in an offline way is make adjustments for the differences between your own and the other person's wants and beliefs. You take on board others' wants and beliefs, holding the rest of your psychology constant, and then "run the simulation" to see what you would do next. To use a famous example from Gordon (1986), suppose you are trying to predict what your friend would do on hearing footsteps in the basement.

> I imagine, for instance, a lone modification of the actual world: the sound of footsteps from the basement. Then I ask, in effect, "What shall I do now?" And I answer with a declaration of immediate intention, "I shall now . . ." This too is only feigned. (Gordon 1986, 161)

The trouble is that "simulation," as conceived in the cognitive science tradition, is fundamentally a predictive device. This offline approach cannot distinguish between causal pathways and mere correlations. That is, in the simulation, you take on the belief that you can hear footsteps in your basement. And perhaps you find next that within the simulation, you are nervous. It may be that the footsteps cause you to be nervous. But it may also be—though this would be a slightly unusual case—that it is only when you are nervous that you ever pay enough attention to what is going on in the basement to hear footsteps there; perhaps the footsteps add nothing to your original nervousness. In either case, when you simulate the hearing of footsteps in the basement, you will, offline, be nervous. But the simulation will not of itself tell you which causal route is operating to generate the nervousness—whether the footsteps cause the nervousness or the nervousness is a precondition of hearing the footsteps.

This kind of point is graphically illustrated by the "two–system" model of fear by LeDoux and Pine (2016). Suppose you encounter something dangerous. You feel afraid, and then you run. That natural model is that perception of the thing caused you to feel fear, and that feeling of fear was what propelled you out of the area. As LeDoux and Pine put it, the "Fear Center" model is the natural model (see fig. 2.6). According to LeDoux and Pine, though, this model does not really fit the data. Feelings of fear are not well correlated with defensive behaviors or physiological responses; subliminal perception of threatening stimuli generates defensive physiological and behavioral responses in the absence of any feeling of fear; blindsight patients respond defensively to threatening stimuli without reporting any feeling of fear; and although damage to the amygdala interferes with the production of physiological and behavioral responses to threatening stimuli, the feeling of fear is still produced. Taken together, these points suggest that there are actually two systems in play here: one, nonconscious, running through the amygdala, generates the physiological and behavioral responses; the other runs from the sensory system through to the cortex and generates the feeling of fear (see fig. 2.7). Even if there are two different systems here, there must be many connections between them; for example, the defensive survival circuits may modulate the experience of fear.

The distinction between a one-system and a two-system view of the fear-anxiety circuit matters when we consider the use of animal models to test potential treatments for related disorders. It's proven difficult to get effective treatments by this route, and

A. The "Fear Center" Model

Threat → **Sensory system** → Fear Circuit (*Fear*) → **Fear responses**
Defensive behavior
Physiological responses

Figure 2.6. The "Fear-Center" Model. (Joseph E. LeDoux and Daniel S. Pine, "Using Neuroscience to Help Understand Fear and Anxiety: A Two-System Framework," *American Journal of Psychiatry* 173:11 (2016), 1083–93, figure 1A, © 2016, by permission of American Psychiatric Association. All rights reserved.)

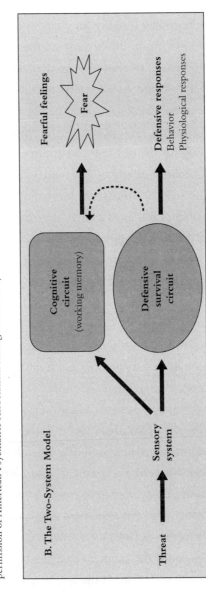

B. The Two-System Model

Threat → **Sensory system** → Cognitive circuit (working memory) / Defensive survival circuit → **Fearful feelings** (*Fear*) and **Defensive responses**
Behavior
Physiological responses

Figure 2.7. The Two-System Model. (Joseph E. LeDoux, and Daniel S. Pine, "Using Neuroscience to Help Understand Fear and Anxiety: A Two-System Framework," *American Journal of Psychiatry* 173:11 (2016), 1083–93, figure 1B, © 2016, by permission of American Psychiatric Association. All rights reserved.)

the two-system analysis suggests a reason why. The animal models are tested by looking for changes in the animal's defensive responses to stimuli. But in humans, the efficacy of treatment is tested by looking for amelioration of the feelings of fear and anxiety. If there is just one fear-anxiety circuit, then this approach makes perfect sense. Suppose a treatment for fear, such as threat extinction (repeated presentation of the threatening stimulus without anything bad happening), is administered. In animals, the reduction of fear is exhibited principally by looking at the defensive responses of the animal. In humans, we look for amelioration of the feeling of fear as exhibited in verbal report. However, the two-system analysis suggests that a treatment that works well in the animal models may be addressing a system that does not implicate the feeling of fear at all: the circuit, remote from consciousness, that generates defensive behaviors. The system that is being monitored in humans is the system that generates feelings of fear or anxiety and exhibits those feelings in verbal behavior. So it is perhaps not all that surprising that treatments that work well on one system are not having much impact on the other system. That's not to say that previous work will be of no help with human disorders. Humans may have disorders of either or both of those circuits. And animal work may well help when humans have problems with the defensive-behavior circuit. To address the feelings of fear or anxiety, however, a different approach is required. Here, it seems possible that clinical work, perhaps using cognitive-behavioral therapy, will be more valuable.

Suppose for the moment that the two-systems view is correct. Suppose that, not knowing any of this data, you try to simulate the condition of someone feeling fear. Well, you would likely only feel the fear if you perceive a threatening stimulus. And if you perceive a threatening stimulus, then your nonconscious fear circuit likely would activate, so you might be engaging in defensive behaviors. When you simulate offline the feeling of fear and run the simulation, you will get the upshot that in that case, you will be engaging in defensive behavior. But the fact that the simulation, when executed correctly, delivers that upshot does nothing to suggest that the feeling of fear caused the defensive behavior. That was not the point or promise of the exercise of simulation. The upshot was not a causal hypothesis but merely a prediction that when feeling the fear, you will also engage in defensive behavior. The correctness of that prediction could be underpinned by any of a wide variety of causal structures, and the simulation exercise of itself is not an attempt to specify which causal structure is in question here.

Simulation theory, as usually conceived, is not even an attempt to characterize our knowledge of singular mental causation. It is entirely a predictive exercise and makes no attempt to characterize our knowledge of the one-off, idiosyncratic causal processes that may engage us in a court of law—for example, in finding why the defendant acted as they did. Therefore, it does not provide a characterization of the imaginative understanding we can and do use in the law courts and in everyday life.

Simulation theory is usually contrasted with "theory-theory" approaches to other minds (e.g., Gopnik and Meltzoff 1997) and with the classical argument from analogy. But the theory-theory is explicitly a matter of finding generalizations about the mind based on oneself and the case of others. Theory-theorizing does not provide any way at all of establishing singular causal claims.

Similarly, using the argument from analogy is a matter of establishing psychological generalizations on the basis of one's own case and extending them to the case of other people. It provides only a way of discovering generalizations that might be applied to both oneself and other people. It gives no way at all of finding the truth of singular causal claims.

But in practice, we do think that our imaginative understanding of one another, following the trajectories of one another's thoughts and feelings, provides us with knowledge of singular causation. And as in the case of Othello, our acquaintance in the elevator, or the defendant in court, we give this imaginatively grounded knowledge of singular causation great practical weight.

It is true that philosophers have sometimes written scathingly about the idea that causal relations can be discerned by means of imaginative understanding. It is also true that imaginative understanding requires a lot of empirical control; as Jaspers said in a passage cited above, we require "*the tangible facts* (that is . . . the verbal contents, cultural factors, people's acts, ways of life, and expressive gestures)" to ground an imaginative understanding of singular causation. Much of Freud's work, for example, can be seen as an attempt to generate an imaginative understanding of singular causation in patients beginning from quite distant origins, and the speculations are generally not grounded in tangible facts; for example, in the case of "rat man," the idea is that one can follow the train of thought and feeling from an early traumatic hearing of a gruesome story through to a complex of present-day behaviors (Freud 1909). The vindication of the story was thought to be in the success of the therapy used on the strength of the story. This evidently leaves Freud exposed to a charge that he is simply making up the whole thing. Grunbaum (1990), for example, claims that there is nothing more to the use of imaginative understanding than the detection of "thematic affinities" between various psychological states of the patient, and he points out that this is not an adequate basis for knowledge of causation.

> Narratives replete with mere hermeneutic elucidations of thematic affinities are explanatorily sterile or bankrupt; at best, they have literary and reportorial value; at worst, they are mere cock-and-bull stories. (575–576)

Although one might sympathize with Grunbaum's reservations here, the natural question is, If we do not rely on our imaginative understanding of one another to generate knowledge of singular causal connections in the mind, how do we ever get knowledge of singular psychological causation? As we've seen, we do not achieve such knowledge by the use of massive experimentation and observation; we do not do it by tracing neural pathways. So how do we achieve the knowledge of singular psychological causation that matters so much in everyday life? Not by simulation, not by the generalizations of an argument from analogy, and not by articulating a theory of human behavior. Grunbaum has no answer to this question. The fact is that he has mistaken a problem of relative detail in Freud's approach—that his hypotheses are not sufficiently grounded in "tangible facts"—for the diagnosis that there is something in principle wrong with the whole idea of achieving knowledge of singular causation by means of imaginative understanding. The fact is that our ordinary imaginative understanding of one another is often well-grounded in the tangible facts. Suppose we go back to the defendant who is claiming that he acted out of a desire to defend himself rather than out of a desire for revenge. Suppose you are on the jury and are trying to achieve an imaginative understanding of how the attack was generated by his frame of mind. Now suppose there is a witness who testifies that just as the knife went in, the defendant said, "That's for Billy." In the right context, this tangible fact can be absolutely conclusive in establishing that one imaginative understanding of the defendant rather than another was correct, and it can justify a finding of guilt beyond reasonable doubt. It is not as though this possibility of grounding in tangible fact is limited to relatively brief and simple exercises of imaginative understanding. We'll end this chapter with a characteristically spectacular exercise by Sherlock Holmes; the point here is to note how well grounded in the tangible facts is his exercise of imagination in following the generation of a train of thought. Watson tells us what happened.

> I had tossed aside the barren paper, and leaning back in my chair, I fell into a brown study. Suddenly my companion's voice broke in upon my thoughts.
>
> "You are right, Watson," said he. "It does seem a very preposterous way of settling a dispute."
>
> "Most preposterous!" I exclaimed, and then, suddenly realizing how he had echoed the inmost thought of my soul, I sat up in my chair and stared at him in blank amazement. . . .
>
> "Do you mean to say that you read my train of thoughts from my features?"
>
> "Your features, and especially your eyes. Perhaps you cannot yourself recall how your reverie commenced?"
>
> "No, I cannot."
>
> "Then I will tell you. After throwing down your paper, which was the action which drew my attention to you, you sat for half a minute with a vacant expression. Then your eyes fixed themselves upon your newly-framed picture of General

Gordon, and I saw by the alteration in your face that a train of thought had been started. But it did not lead very far. Your eyes turned across to the unframed portrait of Henry Ward Beecher which stands upon the top of your books. You then glanced up at the wall, and of course your meaning was obvious. You were thinking that if the portrait were framed it would just cover that bare space and correspond with Gordon's picture over there."

"You have followed me wonderfully!" I exclaimed.
"So far I could hardly have gone astray. But now your thoughts went back to Beecher, and you looked hard across as if you were studying the character in his features. Then your eyes ceased to pucker, but you continued to look across, and your face was thoughtful. You were recalling the incidents of Beecher's career. I was well aware that you could not do this without thinking of the mission which he undertook on behalf of the North at the time of the Civil War, for I remember you expressing your passionate indignation at the way in which he was received by the more turbulent of our people. You felt so strongly about it that I knew you could not think of Beecher without thinking of that also. When a moment later I saw your eyes wander away from the picture, I suspected that your mind had now turned to the Civil War, and when I observed that your lips set, your eyes sparkled, and your hands clinched, I was positive that you were indeed thinking of the gallantry which was shown by both sides in that desperate struggle. But then, again, your face grew sadder; you shook your head. You were dwelling upon the sadness and horror and useless waste of life. Your hand stole towards your own old wound, and a smile quivered on your lips, which showed me that the ridiculous

side of this method of settling international questions had forced itself upon your mind. At this point I agreed with you that it was preposterous and was glad to find that all my deductions had been correct." (Conan Doyle 1917, 66–67)

The key point here is how epistemically well grounded is Holmes's imaginative understanding of the causal progression of Watson's thoughts and feeling. It's cautious, with each conjecture along the way being tested against the evidence of Watson's direction of gaze and behavioral expression. And it's hard to see how Watson's reaction to the final remark about what a preposterous way this is of settling a dispute could be explained otherwise than as vindicating Holmes's reconstruction.

CHAPTER THREE

SOCIAL ROBOTS

I. SOCIAL ROBOTS

I was introduced to social robots some years ago by finding myself holding one of Alex Reben's BlabDroids. They're made of cardboard, are about the size and shape of a pizza box, and roll around on tracks. They have a smiley face crudely felt-penned onto them. One of the eyes is a video camera. They speak in the voice of a seven-year-old child. They roam around venues like conference centers, hotel lobbies, or art galleries, and they ask people questions like, "What's the worst thing you've ever done?" or "Who do you love the most?" Reben's group first realized they were on to something when a runner in the Boston marathon briefly timed out in the conference center, then on being encountered by the BlabDroid lay down on the floor beside it and talked to it for about ninety minutes, pouring out the most profound details of his life. Of course, people don't always respond so strongly, but many do. Their reactions to questions like, "What's the worst thing you've ever done?" or "Who do you love the most?" can be astonishingly full, not to say overwrought and candid. A common reaction is to pick up the BlabDroid and say, "I can't talk here with all these people around, but come into the corner and I'll tell you about it all." There are a number of differences between the ways people respond to BlabDroids and the ways they respond to other humans. They often seem to be more honest and forthcoming with the BlabDroids than they would with another person. One possible use for them is in police interviews because people feel some freedom from the judgments of another human.

Social robots in one form or another have been used for decades in work with autistic children. They allow children to have free practice in basic types of social interaction without putting the cost on human carers. They have also been used for decades in caring for the elderly and lonely, who may not even realize that what they are dealing with is a robot. If your parent is dying alone on the other side of the country, there is a question whether you would help the person better by using your money for sporadic and infrequent trips to visit, or in investing in a robot that will talk to the parent, never forget anything the person says, be tireless in its interest, and send you videos of the interactions. As with all technology, the development of social robots is proceeding in leaps and bounds.

So is the demand. Many countries have a boom in social isolation as more people, particularly young people, live in single households. If the demands of marriage and family are too much in a pressured work environment, social robots are designed to alleviate the difficulties in living alone.

There are two aspects to social robots. One is the software. We have programs that can engage in simple linguistic communication, remember information they're given, and connect with other robots and other computers. The other aspect is the hardware. People in robotics and animation sometimes talk about an un-canny valley: things that are clearly not human can draw cuteness or other positive responses, but as they get more and more like humanlike, but still subtly not human, they seem eerie and spooky. Then as you get something indistinguishable from a human, you get an ordinary response (Mori 2005). Alex's BlabDroids are on the far side of the uncanny valley. Yet they draw very strong emo-tional reactions from people. The hardware seems to be an im-portant part of this. Just as when you meet someone for the first time, before a word has been exchanged, there is a lot going on with bodily organization, how you adapt your body to the pos-ture of the other person, whether you seem detached or engaged, and so on. The hardware of a suitably configured robot, which-ever side of uncanny valley it's on, can be doing a lot of work to make possible that groundwork for emotional connection with a human (cf. Philips et al. 2018). For example, if a robot is going to work with you, if it is going to hand items back and forth with you, then it has to be built so that there is a perception of compe-tence, warmth, and comfort (cf. Pan, Croft, and Niemeyer 2018). The effect of a well-designed package of software and hardware in a social robot could be intensely engaging emotionally. And it's not just strong emotions in the way that a sunset might draw a big response from you. The complex responses Alex Reben's simple robots draw—trust, confessions, and so on—go well beyond that.

Once I'd seen Alex Reben's BlabDroids, I gave a number of talks with him about them to quite varied audiences. I was struck by the occasionally passionate reactions the topic produced: a com-mon reaction is that there is something awful, sad, or pathetic about this phenomenon, though people usually had difficulty explaining what. If the robots were merely seen as amusing adjuncts to an ordinary social or domestic life, people had much less difficulty thinking of them as possibly life-enhancing and welcome. But the negative reaction is one that we will have to grapple with in many areas. Suppose you have to choose between sending your child to a school where second languages are taught by regular human teachers, and one where the bulk of second language teaching is done by robots. The academic outcomes at the two schools are broadly similar. Is there any presumption in favor of the human school? Is there anything better about interacting with humans than interacting with robots?

One model for the role of social robots in alleviating social isolation is provided by George Eliot's *Silas Marner* (Eliot 1861). Silas is a lonely and miserable miser who has a cache of gold that he loves very much. It's a strong, unidirectional emotional relationship between him and the gold, perhaps like the strong, unidirectional relationships we'll be having with our robots. And then someone steals the gold. Silas is disconsolate and virtually loses his reason after the loss. Then a single mother with a newborn baby abandons the baby on Silas's doorstep. Silas opens the door and sees only the golden head of hair. At first, in his distraught state, he thinks it's his gold come back to him, but as he plunges his hand toward it, he finds himself picking up the child. Having found the child, he takes her in and reluctantly takes responsibility for her. Having done so, his life is transformed: the village, observing the lonely old miser trying to tend to the child, rallies round, draws him to their bosoms, and helps him look after her. One way of articulating the problem is to say that if it's a person that's reading you to sleep, that person is going to be socially connected in a way that a grinning machine cannot be.

The trouble with that model for our questions about social robots is that social robots themselves are likely to be widely, endlessly socially connected to real people in a way that a crock of gold cannot be, and indeed, in a way that an ordinary person could not be without the benefit of the Internet. At the simplest level, with the BlabDroids, although they're just cardboard boxes, there is a person behind them; there is someone—or maybe a lot of people—viewing the tapes and able to respond to them. However, there is maybe a background concern here that this kind of remote, mediated connection to other people is going to feed into a general attenuation of our social connections with one another. And indeed that can be generalized. For example, if you have a number of single households, each with its own social robot, they can be connected to one another. They may, for example, be sensitive to the background noise in the house, and they may be able to detect when there is any problem in their own home and communicate it to the others. Someone living alone with a social robot may have wider connections to other people, each similarly living alone with a social robot, than any purely human set of mediators to a village could provide (cf. Jeong et al. 2018).

Proponents of social robots often use the analogy with dogs. The dog is unarguably a highly successful piece of engineered biological social tech, devised to provide us with a dimension of companionship that we wouldn't usually have, and it is generally regarded as unquestionably benevolent. Yet we do regard interaction with dogs as not as valuable as interaction with other people, and the question is why, given that we usually don't regard our interactions with dogs as unpredictable or dogs as not dependable. In Chekhov's story "Misery" (Chekhov 1886), the coachman's son has died, and the coachman is understandably very upset about this. He tries to tell one of the passengers about it but is dismissed. One by one, he tries to tell the other passengers about his grief, and no one wants to know. He tries to tell the stable boy, but he won't listen either. Ultimately he tells the horse, who whinnies gently. The problem here is not that the horse's response is unpredictable or malevolent, but this is certainly a picture of misery. But why is that? Why would it be better to be talking to a human? And would it be any better if he could explain the thing to a social robot?

One way to put the question is to ask, "Which activities are we glad to have automated, and which not?" (One of Alex Reben's constructions is a hand-operated mangle for popping large quantities of bubble-wrap fast; he's considering a motorized version.) Suppose it's the 1950s. You and your spouse have worked out a division of household chores whereby on the weekend, you wash the car together. In practice, though, neither of you is wild about the task. Then a carwash is built in your town. A machine can now, to your relief, do the task you used to do together.

There must be thousands of couples who used to drive together regularly, with one driving and the other navigating. This is a big way in which people used to communicate. Now, with GPS car navigation systems, this line of communication between couples has been lost, to the great relief of most people. Here, the automation seems unquestionably a good thing despite the loss of emotional contact between real people.

Suppose now it's 2030. Your spouse often arrives home full of the perplexities of the day from her pressured career and will often spend the evening telling you about them. You're glad to know, to be involved in her life, and to offer comments where you can. But it's sometimes an effort to focus. You often tell her the interesting problems from your day too, though sometimes she finds it hard to keep track. Then you each get a social robot. It makes eye contact, asks you each about your day, and provides soothing and supportive reactions. Moreover, it never forgets; everything said is recorded and cross-indexed by the system. You now unload the day on the robot rather than on each other. Isn't this just as good as the carwash?

To set out the general question schematically, let's consider two people, Ajay and Bebe. Suppose that Ajay has an ongoing emotional reaction to machine M1, and Bebe has an ongoing emotional reaction to machine M2. Let's suppose that M1 elicits the same emotional reaction from Ajay that Bebe would. And suppose that M2 elicits the same emotional reaction from Bebe that Ajay would. But neither M1 nor M2 can credibly be supposed to have anything in the way of thoughts or feelings (see fig. 3.1). Now consider the contrast between that picture and the one in figure 3.2. When it's stated at this general level, it seems to me that people often feel that what's going on in this second picture is better than what's going on in the first picture. It's better if we have a single relationship between two people than if we have two separate unidirectional relationships between a person and a machine. And that's so even if the level of contentment of the people involved in the two-way relationship is exactly the same as the impact on the people involved of the two unidirectional relationships. We don't want our emotional lives to be automated. Although I myself feel this reaction very strongly, it's hard to defend. It's hard to explain why, and in what sense, it's better. But

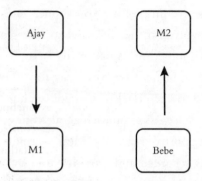

Figure 3.1. Ajay and Bebe are ordinary humans; M1 and M2 are machines interacting with them. In context, Ajay's interaction with M1 produces in Ajay the same or better levels of satisfaction than would Ajay's interaction with Bebe; similarly, Bebe's interaction with M2 produces in Bebe the same or better levels of satisfaction than would Bebe's interaction with Ajay.

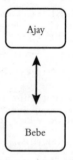

Figure 3.2. Direct social interaction between Ajay and Bebe.

this reaction is going to come under a great deal of pressure as we encounter robots that are going to play with and teach our children, provide lonely people with friendship, and give comfort to the elderly and confused. Of course, it's easy to say that what a robot provides may be better than nothing. But as clever and ingenious people continue to develop these things, the question whether what the robot provides is just as good or better as what one could provide oneself will become ever more pressing.

2. CONSCIOUSNESS

Here's a remark by Bertrand Russell on one of the reasons why he sought love all his life.

> I have sought it, next, because it relieves loneliness—that terrible loneliness in which one shivering consciousness looks over the rim of the world into the cold unfathomable lifeless abyss. (Russell 1967, 9)

Suppose you are looking over the rim of the world and into the cold, unfathomable, lifeless abyss. You find there a social robot. What would the robot have to be like for it to help? Or let's put it another way. We are naturally highly defensive about admitting loneliness. We want to press on regardless, dismissing such concerns, and with a bit of luck, other people will take a shine to us. Going on about one's own loneliness is likely to be unhelpful in finding other people to alleviate it. Suppose your father is achingly lonely. He lives on the other side of the country, in an apartment on his own. His only visitors are a rotating cast of paid helpers. You visit when you can, but those visits every three months, though appreciated, do not really address the problem. You have the money to buy him a social robot, but would it help? What would it have to be like for it to address the core problem, the loneliness, as opposed to merely providing some agreeable diversions?

A natural thought is that the robot isn't conscious. After all, we are talking about relatively simple robots rather than the things you find in science fiction movies, where there is serious uncertainty as to whether the machine is sentient. But what exactly are we talking about when we talk of consciousness? Philosophy of mind in the last fifty or sixty years has focused on the idea that there is some qualitative aspect to consciousness that is peculiarly difficult to explain in computational or physical terms. To illustrate, there is the problem of the inverted spectrum. It's often been said that two people could be, for all practical purposes, computationally and physically identical, but maybe when one looks at a violet, that person has the same color sensations that the other one has when he or she looks at a marigold, and so on. Those color sensations thus can't be analyzed in computational or physical terms. Presumably the robot lacks such sensations, and perhaps that's why we feel qualms about the idea that the robot can relieve loneliness.

The question is about what matters in the having of a mind. Mentality is complex. One way of focusing the question is to ask what would matter to us, what would be important to be there in a social robot, if it were to be capable of relieving loneliness and if it were to provide a mind with which one could make contact. The answer suggested by the intense focus on qualia—those qualitative aspects of experience—in the last fifty or sixty years is that it's having the qualia that is the critical thing. But why should that be the key thing? Suppose there are these qualitative aspects to experience. Why would they matter to us socially? Suppose there were a robot that was computationally and functionally similar to a human but didn't have qualia. Would that be an inadequate substitute for a human? It might turn out that some humans don't have qualia, just as it has been suggested that some humans really do have spectrums that are inversions of the usual color sensations (Nida-Rümelin 1996). Would these people therefore be inadequate as social companions? Why should we attach such a high value to qualia?

There are views on which the foundation of all ethics is the positive value of sensations of happiness and the negative value of sensations of pain. These views are more often encountered in introductory philosophical sessions rather than in developed academic work, but they are recognizable (cf. Darwall 1974 for discussion of sophisticated variations on this idea). If that's right, then there would be nothing to value in the mental life of a robot without qualia.

This is an engaging idea, but it doesn't seem compelling. After all, the very idea of qualia is also one of the most heavily criticized in philosophy of mind. Consider a simple, radical position on which there are no such things as qualia. Perhaps qualitative colors are characteristics of physical things rather than characteristics of one's mental life. This has some claim to be the view of unreconstructed common sense. Colors seem to be out there, on the object. Experiences of colors are experiences of those external characteristics. People who know a bit of science, such as Locke (1690 / 1975) or Boghossian and Velleman (1989), sometimes argue that science has shown our ordinary view of color to be mistaken. This "error theory" of our ordinary conception of color presumes the intelligibility of the idea that the colors are out there, on the objects—that's the view that science is thought to overturn. But suppose now that science does not overturn the common-sense view, as some philosophers have argued (Moore 1903; Allen 2016; Campbell 2020). Suppose there are no internal qualia of color. Color experience, on this view, is simply a matter of being experientially related to the qualitative characters of the physical objects around us. Now, there are a number of objections you might raise to this common-sense view, but it does not seem right to argue that this common-sense view somehow threatens the value of ordinary human society. After all, human society seems to have the value it does even to those innocent of physics and the internalization of color experience.

One might argue that even though it's possible to resist the analysis of color experience in terms of the colorlike characteristics of inner sensations, the key phenomena for social value are happiness and pain, which are nothing but inner sensations. But even this can be resisted. One might argue, with Foot (2001), that happiness, properly understood, is "enjoyment of good things," where a "good thing" is something external, such as "feeding a lot of people," or "getting the plant properly dug in." "Enjoyment of" the good thing is the relation one stands in to it, and that's what happiness is. There is no inner quale of happiness in this view: you can challenge someone's claim to be happy, for example, by challenging whether what they are enjoying really is a good thing. Similarly, pains might be thought of as relations the subject has to external phenomena (external to the mind, that is). After all, the pain is in one's hand or one's tooth; it's not in the mind. One could think of the experience of pain as the relation one stands in to that phenomenon in the hand, or in the tooth. Again, these views of happiness and of pain have much to be said about them. But there seems to be no force to the idea that these views would overturn our ordinary valuation of social interaction with other people.

If we are to contrast the social value of interaction with a person with the social value of interaction with a social robot, it doesn't seem that it can lie in the fact that the person has qualia whereas the social robot doesn't. Even if people don't have qualia, there is still social value in interaction with them. We still have to understand what would have to be present in a social robot for interaction with it to have the same kind of value that there is in interaction with a person.

You might say, Well, perhaps the key thing is not properly thought of in terms of qualia, but it still has to do with experience. Even if "qualitative character" is out there in the world, the important thing is that there should be experience of it. So what matters, and what is missing in the case of the robot, is an experiential point of view from which the world is being encountered (Nagel 1986). This concept of the experiential point of view is, like the idea of qualia, one that is often thought to be peculiarly resistant to analysis in computational or physical terms (cf. Nagel 1971). Presumably, it will be particularly difficult to build a robot that has an experiential point of view on the world. And perhaps that is what is required to make a robot that would have the social value of a human being.

The trouble with this line of thought is that as Nagel (1971) pointed out, there are humans who do not have a single point of view on the world. Patients who have had a cerebral commissurotomy to combat epileptic seizures exhibit a number of puzzling phenomena in virtue of which it's hard to say that they do have a single perspective on their environments. Yet those patients seem to be as socially valuable as anyone else. This point was poignantly impressed on me in an undergraduate class in which I was explaining to the students that split-brain patients are, in ordinary life, functionally indistinguishable from other people. One student put up her hand and said she'd heard that these patients were often socially dysfunctional. After she'd finished speaking, another student put up her hand and said that she herself was a split-brain patient, she'd had the procedure on account of epilepsy, and she was perfectly socially functional, thank you. From the bravery, lucidity, and grace of her performance, it was quite obvious that she was right. From a couple of minutes spent looking at the readily available videos of split-brain patients, it's evident that these people are of as much social value as anyone else. Therefore whatever is missing in a social robot, whatever the component that would explain the difference in social value between such a machine and a human, it does not seem that it can be the presence of a single experiential point of view.

To sum up, the idea that what is missing in a social robot is consciousness is not of much help. The current interpretations of what it would take to have a mind in terms of (a) qualitative character and (b) unity of consciousness do not seem to give much of a fix on what is required.

There is a simpler point to make. Suppose we go back to the Chekhov story, "Misery." The horse to which the coachman pours out his tormented story is conscious. If there are qualia and unified points of view on the world, the horse has those. Yet the coachman pouring out his heart to the horse does not seem to be in any better a position than someone pouring one's heart out to a social robot. The position is different in that the social robot can be sensitive to the language that the person is using, and it can give responses in language. But it is hard to feel that Chekhov's coachman is in a better position than the user of a social robot. Therefore the presence or absence of qualia and a unified point of view cannot be the key things.

I think we get a better start on this by appealing to concept of empathy. What is awful about the idea of your aged parent explaining their loneliness to a social robot is that the robot lacks any capacity for empathy. It matters to us that we are empathized with, though of course there is a question regarding what concept of empathy is relevant here. There is an immediate issue whether this can give us a comprehensive answer to the question of what we'd be missing in interactions with a social robot. Wouldn't there be more missing than that? In particular, there's an issue that arises because this is a machine: even if it could provide empathy, wouldn't there be an aspect of freedom or agency that is missing here? Isn't part of what is wrong with the idea of social interaction that this is something designed, bought, and paid for that can be switched on and off by the user at will? This is indeed a significant issue and raises the question of what concept of agency is important. We'll come back to this; for the moment, I want to focus on the empathy itself rather than the questions of whether it's freely given and why that would matter.

The more pressing question is what concept of empathy matters here. After all, you might say that an appeal to empathy is not so very different from an appeal to conscious experience. Empathy is sometimes thought of as a matter of there being a kind of mirroring of the qualitative aspects of experience. You are looking at a red thing, so your experience has a kind of red quality; empathy is a matter of my having a red experience too, perhaps in imagination rather than perception, but nonetheless it is an experience that has that "red" qualitative character. Or you might argue that empathy is a matter of knowing how things are from the point of view of another person. That is, you are thinking of the other person's experiences as having a kind of unity in that they can all be surveyed from a single point of view. Empathy is when, in imagination, you can survey how things are from that point of view. The robot, not having a conscious point of view itself, isn't in any position to survey how things are from your point of view. Now, in this picture, qualia and the presence of a single experiential point of view would matter instrumentally rather than in and of themselves. They would matter only because they make empathy possible.

I think that this kind of response leaves the initial phenomenon as puzzling as before. Suppose the thing can manage a kind of snapshot reflecting your qualitative state. Why does that matter? Suppose the spectrum of color experience of those you know is an inversion of yours, so they never manage to mirror your qualitative state. Would that mean that social interaction with them is no more valuable than social interaction with a social robot? Or suppose again you're dealing with split-brain patients. Would that mean their understanding of your perplexities is of no more value than interaction with a social robot? Whatever the key thing is, it's hard to see what would be missing that's important. Perhaps qualia do make possible a kind of static mirroring of one's qualitative life, but why does that matter?

I think the important things here emerge only when we consider our imaginative understanding of one another as something dynamic, something that relates to how we know about the causation in one another's mental lives. This seems to be the key thing that would be missing in our interactions with a social robot. But to make this fully explicit, we need to look at some basic points about mental causation and our knowledge of it, to which we now turn.

3. SHALLOW VS. DEEP CAUSAL SYSTEMS: EMPATHY AS KNOWLEDGE OF SINGULAR CAUSATION IN THE MENTAL

To address this properly, we need to look a little further at the causal structure of the human mind. What's the relation between general and singular causal claims? One natural idea is that singular causal claims are true because they report instantiations of general causal claims. Recall cellular automata, such as Conway's Game of Life. An example of a cellular automaton is a two-dimensional grid of cells, each of which can be in one of two states, alive or dead. We have a discrete set of times for the system. In the Game of Life, for instance, what happens to any individual cell at time $t+1$ is determined by the following fundamental rules (fundamental in that they aren't derived from any more basic rules governing the system).

    a. If the cell is dead at t, it becomes alive at $t+1$ only if exactly three of its neighbors (out of the adjoining eight cells) were alive at t.

    b. If the cell is alive at t, it stays alive at $t+1$ only if two or three of its neighbors were alive at t; otherwise, it dies.

Suppose we have a particular cell, alpha, which is dead at t. Exactly three of its neighbors are alive. Therefore, the cell quickens into life at $t+1$. We have the causal generalizations (a) and (b) above. Now consider the singular causal question: What caused that cell, alpha, to become alive at $t+1$? Well, we can read it off from the rules: those three cells adjacent to alpha being alive, and their being the only neighbors of alpha alive, caused alpha to become live at $t+1$.

This system is causally shallow in that the only true singular causal claims that apply to it are those that simply instantiate the general laws governing the system. The only conception that we have here of singular causation is instantiation of general laws. I have stated the laws as being deterministic and exceptionless, but of course it could also happen that the laws are probabilistic. It could be that instead of rules (a) and (b) above, the system is governed by two rules:

> a′. If the cell is dead at t, there is an 80 percent chance of it becoming alive at $t+1$ if and only if exactly three of its neighbors (out of the adjoining eight cells) were alive at t. Otherwise, there is an 80 percent chance of it staying dead.
>
> b′. If the cell is alive at t, there is an 80 percent chance it stays alive at $t+1$ if and only if two or three of its neighbors were alive at t; otherwise, there is an 80 percent chance of it dying.

In this system, we can still talk about general and singular causation. The two rules I've just given, (a′) and (b′), state the generalizations governing the system. If we have a cell, beta, that is dead at t and has three neighbors that are alive, then beta quickens into life at $t+1$. We can say that those three neighbors (and only them) being alive is what caused beta to quicken into life. Of course, if beta doesn't have exactly three neighbors alive at t, there is still a possibility that it will nonetheless quicken into life at $t+1$, but in that case, there is no singular cause of beta quickening into life; it simply happened, and that's part of the system being probabilistic: sometimes there are events with no singular cause.

As I said, the rules I'm describing, (a), (b), (a′), and (b′), are fundamental in that there aren't any more basic facts about the system in virtue of which they hold. If you model a cellular automaton on a computer, there will be more basic facts about the operation of the computer hardware in virtue of which you see the phenomena you do on screen. But for the cellular automaton that is being modeled onscreen, there are no more basic laws. The automaton is completely defined by the basic rules I just set out; there isn't any more to it than that.

The system I've just described is causally shallow in that where we find singular causation, it is only as an instantiation of general laws. To have knowledge of singular causation in this system, you need only knowledge of the basic laws and the context of any particular event whose causes and effects you are interested in. There isn't any more to the singular causation than that. It is not difficult to see that social robots could treat humans as causally shallow systems in this sense and can already do so to spectacular effect. A famous study found that

> computers' judgments of people's personalities based on their digital footprints are more accurate and valid than judgments made by their close others or acquaintances (friends, family, spouse, colleagues, etc.). . . . people's personalities can be predicted automatically and without involving human social-cognitive skills. . . . computer models need 10, 70, 150, and 300 Likes, respectively, to outperform an average work colleague, cohabitant or friend, family member, and spouse. (Youyou, Kosinski, and Stillwell 2015, 1036–37)

Computers can already do well at diagnosing personality traits such as openness, conscientiousness, extraversion, agreeableness, and neuroticism, on the basis of Facebook likes. They do much better at humans on some dimensions, such as openness. They doubtless miss cues that are easily available to humans in unconscious behavioral responses, and so on. But the volume of data available to social robots will expand, and data analysis will become more sophisticated. Therefore a social robot living with you that is connected to

Facebook, Amazon, and so on and that has a massive amount of data on you and on millions of other humans may be able to predict your further likes and dislikes far better than any human, in a way that seems to indicate an uncanny insight into your psychology. Does ordinary human empathy with you provide anything that a social robot, so wired up, could not provide? There is an old joke about someone who complained that their Amazon recommendations were so much more insightful and perceptive than the clumsy gifts that their partner occasionally bought for them. That was a joke, but it may well be the reality before long that social robots will exhibit far more knowledge of our preferences than other humans can. If we substituted our ordinary human family lives for interactions with such robots, would we be losing anything of value? Here is a way to think of it. Consider chess-playing computers. It used to be that chess-playing computers were very good at chess but used strategies quite different from those used by humans; they used big data to determine how to move rather than the kind of intuitive understanding of the chessboard used by humans. Each of us, from when we are born, is fighting to win love, and we carry on fighting for love through our lives until we become old and bitter and filled with thoughts of revenge. Social robots will also fight for our love through their understanding of our psychologies, but using strategies from big data, which is quite unlike the strategies used by other humans. The question is whether we would lose anything of value if we settled for replacing ordinary human families with robots that win our love using big-data strategies.

If human psychology is a causally shallow system in the sense I have explained, in which singular causal relations are merely instantiations of general causal truths about human psychology, then I find it hard to see what we would be losing, other than an indefensible preference for our own kind. But part of the point of our discussion of singular causation in the mental has been that not all causal systems are causally shallow in this sense. There are causal systems, in particular the human mind, for which singular causation is not simply an instantiation of causal generalizations, and here, singular causation has to be understood in terms of both counterfactual dependence and a conception of process. These are systems the analysis of whose causal functioning requires us to introduce some conception of process or causation as production. We have seen the case for saying that the human mind is of this sort; that was the argument of Chapter 2. It is for such systems that Ellery Eells's remark holds true; he puts the distinction between singular and general causation as a distinction between "token" and "type" causation.

1. very little (if anything) about what happens on the token level can be inferred from type-level probabilistic causal claims, and . . .
2. very little (if anything) about type-level probabilistic causal relations can be inferred from token-level probabilistic causal claims. (Eells 1991, 6)

Eells makes his point in the context of a discussion that analyzes general causal relations in terms of the conditional probabilities relating cause and effect variables, and singular causal relations in terms of the evolution over time of conditional probabilities. The structures of the two analyses are quite different and provide a grounding for his claim about the independence of the two. In effect, Eells's analysis of singular causation in terms of the evolution of probabilities over time will give a way of making more explicit the similarity between physical causal processes, such as the falling of a rock, and mentalistic causal processes, such as Desdemona's train of thought and feeling, or the calculations of an arithmetician. But this kind of analysis evidently does not apply to the operations of a cellular automaton. In the case of a shallow causal system for which we do not have any conception of causal process, token causation is merely an exemplification of type causation, and we have ready translations back and forth between the singular and the general level.

To see the general point here, consider a type of example Eells discusses at length: smoking causes cancer. Let's assume that the general causal claim here is true. It can be true even if no one has ever smoked. It can also be true even if there are lots of smokers in the population, but none of them has ever contracted cancer (it might be, after all, that smoking causes cancer, but that by accident, many diverse pathologies invade and kill the humans in our population before cancer is contracted). What I am making explicit is that the reason for the contrast with cellular automata here is that we have a conception of the process by which smoking causes cancer. We can make sense of the general claim being true, even though there are no corresponding cases of singular causation, because we can make sense of the idea that case by case, there may have been an interruption to the process by which smoking generates cancer. Similarly, the reason we need a conception of singular causation in the mental, which is not merely an instantiation of general causal truths, is that we have a conception of the process by which the outcome is generated. In the case of singular psychological causation, empathy is the name for the process by which we trace a singular causal pathway.

This tracing of singular causal pathways in the psychological is what social robots, as currently envisaged, seem most conspicuously to lack. I said earlier that the question about the value of social robots can be put as the question "What are the limits of automation?" We have now one way of answering that. Automation is a matter of reducing everything to general patterns so a machine can be programmed to execute them. We saw earlier the power that machines executing programs at the general level already have in human life—a power that seems bound to increase. But no amount of this kind of thing will give us what we have and what we want most from other people: the capacity to follow the idiosyncrasies of our train of thoughts and feelings in the individual case.

There is an analogy between the picture I am proposing of the causal structure of the human mind and the causal structure of the balls on a billiard table. Billiard tables are often used in discussions of causation as if they were shallow causal systems, in which singular interactions merely reflect general laws. We do have to recognize that when we have a system governed by dynamical laws—that is, laws governing the development of the system over time—it's natural to suppose that if we consider the total state S1 of the system at one time and the total state S2 of the system at a later time, and the laws demand that S1 be followed by S2, then S1 caused S2. Here, there isn't any appeal to the notion of a process. But as Russell (1912) pointed out, this kind of idea can apply only to global states of a system, and we usually think of causation as a local phenomenon. For example, we think that a cue shot can be what caused the red ball to go into the pocket. But there aren't going to be any local laws to say that cue shots always get the ball to go into the pocket. Although we sometimes talk of the pool table as though it's a relatively closed system, in fact it isn't. There can always be outside interference—for example, the ceiling collapses onto the table immediately after the ball has been hit. Even if we are in a deterministic universe, we would have to consider an initial state encompassing the whole world in order to find a law that implies that the initial state must be followed by the ball going into the pocket. We have to rule out the passing meteors or lightning bolts that might interfere with the ball going into the pocket. But we don't usually think of causation in these global terms. The reason is that we think of causation in terms of local processes, such as the movement of a ball and its collisions with other balls, connecting causes and effects—and we don't think of the causation in terms of exceptionless laws at all.

A billiard table with a number of balls rolling around on it is a textbook example of a chaotic system (Sinai 1970). Variations in the initial condition of the system that are too small to be measurable in practice may make a big difference to the outcome. That is, very small differences in the force or direction of a cue shot, or in the positioning of just one ball—differences too small to be measurable in practice—may make a big difference to the outcome; they may make a difference as to whether the red goes in a pocket, for example. Even for someone who knows the relevant laws and who knows as far as is practicable the relevant facts about the initial positions and movements of the balls on the table and the force and direction of the cue shot, that person may be unable to predict whether or not the red ball will go into a pocket. Nonetheless, once the cue shot is taken and the balls roll, with the red going into a pocket, the causal pathway from the cue shot to the ball going into the pocket is almost childishly easy to trace.

One simply has to follow the cue shot from the initial ball struck through its collisions with other balls, and their collisions with further balls, to find a causal pathway from the cue shot to finally the red ball going into the pocket. What makes this tracing of the causal pathway possible is that our knowledge of causation is not instructed merely by our knowledge of laws. We have the conception of a causal pathway, the process from one event to another. In this case, it is easy to say what the components of the pathway are. We have (a) the trajectory of an individual ball over time, transmitting causal influence from one location to another. That is, within a single ball over time, we have the transmission of causation: it is because of the initial cue shot that there is now, a second later, the same ball going past a particular place at a particular speed. Causation has been transmitted from the initial cue shot to that later place. But we also have (b) the transmission of causation from one ball to another when they collide. Causation can be transmitted from the cue shot to a later placing of the ball struck, to another ball, through the collision. It is because we have that conception of a causal pathway from the initial shot to the dropping of the red into the pocket that we know what caused the red to drop into the pocket, even though we could not have probabilistically predicted that the red would go into the pocket. We have a postdictive knowledge of causation that seems to depend not on our knowledge of laws but rather on our grasp of the conception of a causal pathway, put together from (a) the trajectories of individual balls and (b) collisions between balls. Of course, there is a subterranean level at which the system is law-governed— and indeed, for all I have said, these underlying laws may be deterministic—but the relation between (1) the existence of causal pathways and our knowledge of causal pathways and (2) this underlying level at which we have law-governedness is at best indirect. On the face of it, it may be that the existence of causal pathways and our knowledge of them does not at all depend on the existence of the underlying level at which we have governance by law. For example, there doesn't seem to be an obvious contradiction in the idea of a system in which there are causal pathways and there is knowledge of them, even though there is no underlying level at which the system can be described as strictly law-governed.

The billiard table provides a good model for our knowledge of mental causation. Of course, there may be an underlying level at which the whole system is governed by laws, perhaps even deterministic laws. The brain is, on the face of it, at best a probabilistic system; it's hard to find regularities in brain functioning that are probabilistic, let alone deterministic. But perhaps there will turn out to be a level (perhaps the level of fundamental particles) at which the brain can be described as governed by deterministic laws. It seems quite evident, however, that knowledge of the laws at work here and how they apply to particular events is not what we use to establish mental causation. Suppose you are talking with a close friend about some big decision she's about to make: should she take the post she's just been offered in London? You can follow her train of thought through all the different options and factors weighed. You may indeed have more insight than your friend does into what factors are counting and why. You might not be able to predict the outcome at the start of the discussion, but postdictively, you can trace the causal pathways through your discussion and definitively know just what the causal path was to her decision. That's not to say that you are infallible: it can happen that the discussion is a charade and there is some hidden factor you don't know about that was determining your friend to decide one way rather than another.

Similarly, in the billiard ball case, it can happen that there's an unseen magnet or a bias in the table, unsuspected by that observer, that is making the ball travel in one direction rather than another. But in both cases, it can also happen that the observer does get all the relevant causal factors and has the right to decisively say that she knows what caused the outcome. In the case of the billiard balls, it's fairly straightforward, as we saw, to characterize the notion of causal pathway that we need, even if it will take substantial work to find a characterization that will generalize to other cases of physical causation. In the case of mental causation, however, it is our capacity for empathetic or imaginative understanding of the other person that provides us with our grasp of causal pathways that we need. Someone who knows about the causal trajectory of a person's thoughts and feelings may not even be aware that the subject has a brain. So the knowledge of causal pathways that we have here should not, in the first instance, be described as a knowledge of neural pathways. We need to consider the subject's psychological states as such, whether ephemeral or sustained, and the dynamics of the relations between them as they unfold over time.

A simple example is an ordinary conversation. You typically can't predict what the other person will say in an ordinary conversation; that's one reason conversation is worthwhile. Nonetheless, even though the thing was not predictable, it may be absolutely apparent to you, for each thing your interlocutor said, just why the person said it. Just as in billiards you can follow the path of each ball readily once it's happened, so in conversation you can readily follow the other person's train of thought and feelings once it's been executed, even though in neither case are you able to predict what's going to happen on the basis of some general laws. In billiards, conversation, and human life generally, Hegel's remark holds true: "The owl of Minerva begins its flight only with the falling of the dusk" (Hegel 1820 / 1991, 23).

4. SINGULAR VS. GENERAL CAUSATION IN HUMAN VS. ANIMAL PSYCHOLOGY

I think that the point we have reached does something to explain the sense in which humans are free. It's usually acknowledged that human freedom has something to do with the way in which human psychology is causal. But the causation here is usually thought of at the level of counterfactuals. To be free is to be such that you could have done otherwise, had you chosen to. But what I want to suggest here is that we should think of the causation distinctive of human freedom at the level of process. We can, I think, contrast two types of animal. For one, animal psychology is merely a matter of being governed by general rules, and governance by them is presumably beneficial. The psychology of an animal of this type will have the same causal structure as a cellular automaton, with singular causation being merely instantiation of general causation and no need for a notion of process to explain the causal functioning of the system. Causation in human psychology, in contrast, cannot be regarded as exhausted by the general causal truths applying to the system. There are, of course, general causal truths that apply to human psychology, but they do not exhaust our understanding of causation in human life. Take for example the general truths about the psychological factors that predispose an individual to substance use, such as impulsivity, peer group deviance, or stressors in early life. We might be able to make a lot of progress in developing knowledge of such psychological risk factors. But no one expects that this kind of work will be able to explain exhaustively, for each case, why that individual turned to substance use. That does not of itself mean that there is anything inexplicable here. It may still be that for each individual, there will be the idiosyncratic story as to how that person turned to substance use, and that story may be complete, consistently with there being an incompleteness in the account that we have at the general level. That operation of human psychology at the level of idiosyncratic singular causation is, I am suggesting, distinctive of human freedom.

This point bears on our attitude to social robots. If social robots can be completely understood in terms of the general principles by which they have been programmed, they will not exhibit this kind of idiosyncratic singular causation. They may nonetheless be remarkably complex and exhibit a significant understanding, at the general level, of causation in human psychology.

Robot psychology, if the robots are well designed, will be like the psychology of our first type of animals in that we can expect the robots and these animals to lead lives that are well regulated: they will not fall into the substance use or other self-destructive modes to which humans are subject. There is a certain respect due to such things. Because human psychology is not governed by general rules, beneficial or not, it is highly vulnerable to a descent into an unhelpful disorder and chaos. Therefore self-regulation is important to humans in a way in which it is not important to animals. For animals, the way in which they are designed and the principles governing them will generally mean that they lead well-regulated lives that keep them out of trouble often enough.

Humans are not exhaustively governed by principles of causation in this way, and therefore leading well-regulated lives is very difficult for us. One thing can lead to another in a way that is intelligible, but we have no overall orchestration of the causal relations among individual psychological states. Each of us has to manage this overall coordination and regulation for ourselves, and we often find it hard. There is a clear statement of the point in Kant:

> [Freedom that] is not restrained under certain rules of conditioned employment . . . is the most terrible thing there ever could be. All animal acts are regular, for they take place according to rules that are subjectively necessitated. In the whole of non-free nature, we find an inner, subjectively necessitated *principium,* whereby all actions in that sphere take place according to a rule. But now if we take freedom among men, we find there is no subjectively necessitating *principium* for the regularity of actions . . . If freedom is not restricted by objective rules, the result is much savage disorder. For it is uncertain whether man will not use his powers to destroy himself, and others, and the whole of nature. (Kant 1997, 27, 344)

It's instructive to compare Kant's description of animals as "regular" in their behavior with Frankfurt's well-known description of animals as "wantons" (Frankfurt 1971). The reckless substance user seems to be a "wanton" in a way that the average hedgehog is not. What is right about Frankfurt's description is that the reckless human lacks a capacity for self-governance, the voluntary conformance of behavior to beneficial regularities, and that the animal similarly has no capacity for self-governance. However, this animal isn't "free" because its behavior is entirely governed by generalizations, and those will generally be enough to keep it safe. The human wanton is one step along the road to freedom by having a capacity for eruptions of singular psychological causation that are not grounded in causal psychological generalizations. But for that reason, the human requires a capacity for self-governance in a way that the animal does not.

In the introduction to this book, I set out a puzzle. I said that humans seem to be alone in the animal kingdom regarding the significance that they give to singular causation. There is not much evidence for a grasp even of general causation among animals, but we can see in a schematic kind of way why it might be valuable to an animal to have a grasp of general causal relations and a way of establishing when they're present. There's no evidence at all of animals having any grasp of singular causal relations, yet we not only have this concept but also put it at the center of our moral and practical lives. Why should we do that when animals manage so well without it?

I am suggesting that it is distinctive of human psychology that we have singular causal relations that are not grounded in general causal relations. We need a way of finding out about those singular causal relations, and regulating them, in a way that some other animals have no need or use for such a thing. That is what's provided by our imaginative understanding of one another, our ability to get inside others' minds and follow trains of thought and feeling.

Indeed, it seems arguable that having this kind of imaginative understanding requires that one should have a deep causal structure, in the sense that one is capable of singular causal trains of thought and feeling that are not simply instantiations of more general causal truths. Jaspers also pointed out that the way we ordinarily know about the causal pathways taken by someone's thoughts and feelings is through the use of imagination, or empathy. This is the fundamental role for empathy: to allow us to follow the twists and turns of someone's thinking and feeling.

There is indeed a question as to whether it is possible to explain the concept of a causal pathway in the mentalistic case in some way other than as the mere correlate of empathy. That is, we could say "for subject A, there was a causal pathway from X to Y" simply means an observer could empathetically or imaginatively understand the progression from X to Y. There might be no deeper account to be given of the idea of a mentalistic causal pathway. That would be disappointing, but we have to keep the possibility in mind.

In the physical case, the simplest version of a process is provided by the trajectory of a physical object, such as a boulder crashing down a cliff side. Even this phenomenon is not easy to characterize explicitly: should we think in terms of a spatiotemporally connected sequence of time-slices of rock? Or is some further description required? For example, perhaps we should require that the condition of each time-slice be counterfactually dependent on the condition of the immediately preceding time-slice (cf. Shoemaker 1984).

Suppose we start with something like this for our conception of a mental process. We might think of a mental process as a series of time-slices of psychological states of a single person, each of which is counterfactually dependent on the previous time-slice. One possibility is that the meaning or rationality that we look for in empathetic understanding of another person is merely an instrumental way of finding such a sequence of time-slices of psychological states. You might similarly argue in the physical case that the spatiotemporal continuity of a series of physical time-slices is merely our way of finding a series of time-slices, each of which is counterfactually dependent on the time-slices preceding it. In principle, we can make sense of a single physical thing jumping from place to place, as might happen in a cartoon, so long as we have counterfactual dependence of each time-slice on the immediately preceding time-slice. In the psychological case, we might think that there's a bare possibility of the existence of a single psychological process devoid of meaning, so long as later stages are counterfactually dependent on earlier stages.

It does not seem as though this kind of approach will work, because it seems unlikely that the notion of counterfactual dependence will take the weight that is here put on it. We can't merely say, "Each time slice in the sequence has to be such that had the previous time-slice not had the characteristics it did, this time-slice would not have had the characteristics it did," because we have to allow that an ongoing causal process, whether mental or physical, can be affected by outside influences. We need some more internal connection between how the process is at different times than is provided merely by the idea of counterfactual dependence. In the case of physical objects, that seems to be provided by the idea of spatiotemporal continuity. In the case of psychological processes, it is provided by the conception of meaning.

The problem is that it's not easy to make fully explicit what concept of meaning is required here: that's why it's always tempting to fall back on the idea that our only grasp of it is provided by our capacity for imaginative understanding of one another. One model is suggested by the case of someone who has to do a mathematical calculation: the rationality and goal-directed character of the process adds something to the bare counterfactual dependence of one stage of the calculation on earlier stages. This kind of analysis might apply much more widely to any case in which the subject has some instrumental rationality to work through, such as calculations as what trade or profession to go into, how much to spend on what, and so on.

The trouble is that there are many cases in which we have a meaningful psychological process without instrumental rationality being relevant. Desdemona's falling in love with Othello is like that: it can't credibly be presented as a case involving instrumental rationality, and it would take some work to describe it as a teleological or goal-oriented process at all. Nonetheless, it's evidently a meaningful psychological process, of which imaginative understanding can be achieved.

I think it's instructive here to compare Jaspers's conception of imaginative understanding with Collingwood's conception of historical understanding. Collingwood (1959) is plainly operating with the idea of trying to follow someone's thought processes through various twists and turns, but he is more explicit about why imagination is the right notion here. His point is that a mental process can be understood only by locating it in a space of alternatives and recognizing why one path through that space, rather than another, seemed normatively correct to the subject.

> The historian of philosophy, reading Plato, is trying to know what Plato thought when he expressed himself in certain words. The only way in which he can do this is by thinking it for himself. This, in fact, is what we mean when we speak of "understanding" the words. So the historian of politics of warfare, presented with an account of certain actions done by Julius

Caesar, tries to understand these actions, that is, to discover what thoughts in Caesar's mind determined him to do them. This implies envisaging for himself the situation in which Caesar stood, and thinking for himself what Caesar thought about the situation and the possible ways of dealing with it. The history of thought, and therefore all history, is the re-enactment of past thought in the historian's own mind.

This re-enactment is only accomplished, in the case of Plato and Caesar respectively, so far as the historian brings to bear on the problem all the powers of his own mind and all his knowledge of philosophy and politics. It is not a passive surrender to the spell of another's mind: it is a labor of active and therefore critical thinking. The historian not only re-enacts past thought, he re-enacts it in the context of his own knowledge and therefore, in re-enacting it, criticizes it, forms his own judgement of its value, corrects whatever errors he can discern in it. This criticism of the thought whose history he traces is not something secondary to tracing the history of it. It is an indispensable condition of the historical knowledge itself. Nothing could be a completer error concerning the history of thought than to suppose that the historian as such merely ascertains "what so-and-so thought," leaving it to someone else to decide "whether it was true." All thinking is critical thinking; the thought which re-enacts past thoughts, therefore, criticizes them in re-enacting them. . . .

Suppose, for example, he is reading the Theodosian code, and has before him a certain edict of an emperor. Merely reading the words and being able to translate them does not amount to knowing their historical significance. In order to do that, he must envisage the situation with which the emperor was

trying to deal, and he must envisage it as that emperor envisaged it. Then he must see for himself, just as if the emperor's
situation were his own, how such a situation might be dealt
with; he must see the possible alternatives, and the reasons for
choosing one rather than another; and thus he must go through
the process which the emperor went through in deciding on
this particular course. Thus he is re-enacting in his own mind
the experience of the emperor; and only insofar as he does
this has he any historical knowledge, as distinct from a merely
philological knowledge, of the meaning of the edict. (Collingwood 1959, 253–255)

Notice that in this description of what it takes to follow someone
else's train of thought and feeling, there is no suggestion that what
we are dealing with here is the subsumption of someone else's
thinking under generalizations. One the contrary, one has to be
capable of generating a one-off train of thought that follows the
other person and takes into account the specifics of the person's
physical and social context. That is why a social robot that does
not have the elements of freedom will not be able to engage in
the kind of imaginative understanding we have of one another.

Notice that Collingwood's broader point also bears on the question whether psychological processes can be understood as neural
processes. His point is that understanding others' trains of thought is
not merely a matter of going over things they have said or thoughts
and feeling they have had; it requires critical understanding, sensitivity to the possibility of alternative trains of thought, and seeing
why the path not taken was not taken. It is not straightforward
to find a physical correlate of this kind of mental process. We would
have to find a sequence of physical states that are physically causally

connected—one set of cell firings causing another set of cell firings, and so on—and it would have to be essential to our understanding of this sequence as a causal sequence that we have some sensitivity to the counterfactuals governing other possible firings that did not but could have occurred, as well as some sense that the firings that did happen were happening because of their normative correctness. It is not easy to see how this would go. One can make the claim that a particular set of cell firings realizes a particular representation in the brain, and that another set of cell firings (perhaps caused by the first) realizes another representation in the brain. But if we consider the cell firings merely as cell firings, it is difficult to see how we will find a notion of normative correctness, recognizable at a merely bi-ological level of description, that allows one to do an analog of the Collingwood exercise at the level of brain biology. If we work with Collingwood's picture of what we are doing when we follow some-one else's thought processes, locating it in a space of possible alterna-tives, there is no apparent way in which we will be able to locate the biological sequence in an isomorphic space of alternatives, with one path through the space being followed because of its normative cor-rectness. I do not say that the thing is impossible, only that I do not see how it is to be done. Notice that there is a way of talking about the brain that is very often used by those who know a little bit about it, where you simply translate back and forth freely from psycho-logical terms to neural terms, as when someone says, "That cake shop really stimulates my endorphins, so I'll go there." Here, the talk about endorphins is simply a dummy, a stand-in for ordinary psychological talk. Suppose that the emperor makes the decision that the law courts should close during Holy Week. We can reenact the train of thought leading up to the decision, just as Collingwood describes: weighing the problems the closure will raise against the benefits, and so on. We could also describe a sequence of biological states and claim that each one of these constitutes a realization of one of the Emperor's judgments along the way. We could say that the biological sequence takes the form it does because of what thoughts are being realized, and we could apply Collingwood-style understanding to the sequence of thoughts. But what is hard to see is how the biological sequence, understood at the biological level, could be thought to be a causal process in the same way that the mental process realized is a causal process. Of course there are biological processes, but they do not seem to be answerable to anything like the constraints to which thought processes are answerable. The disordered thoughts of a schizophrenic patient exhibiting formal thought disorder, for example, may be biologically comprehensible in the same way as any other, even though we do not have a psychologically comprehensible causal process here in Jaspers's or Collingwood's senses. There is a sense in which Emperor Theodosian's thought processes constitute a causal process but the thoughts of the disordered patient do not stack together to constitute a causal process, and it is hard to see how the distinction can be biologically grounded.

Still, Collingwood's emphasis on the normative assessment of the sequence in a mental process does not seem to cover all cases. The kind of normative assessment he's describing seems only glancingly related to an imaginative understanding of Desdemona's falling in love with Othello, for example. At the moment, it is an open question whether we can give a single analysis of the concept of a psychological process that will cover all cases, or whether we have only a patchwork of different styles of process.

It also seems possible that a purely individualistic conception of psychological process can't be sustained in the end and that we will have to understand the concept of the psychological processes of an individual as an element in our broader understanding of narrative: our ability to tell comprehensible stories about the interactions between people that "make sense." As with when we're thinking about individual psychological processes, the notion of a narrative structure is something that we each seem to learn through embedment in our communities, picking it up from characteristic patterns of storytelling. But just as with the individualistic case, it may be that our culturally conditioned understanding of patterns of narrative is an attempt to pick up on some independently existing networks of causal connection.

The problem with the project of programming a social robot to be empathetic is that the natural way to approach programming is in terms of generalities: in a situation of type X, do Y. That's what's done by programs in Amazon and Facebook that work on the basis of what you've bought or what you've liked to make predictions about you. It is much harder to see how to program a computer so that it engages with the genuinely singular. It is not at all straightforward even to give a computational account of our ability to visually keep track of the trajectory of an ordinary physical object. Some theorists, such as Zenon Pylyshyn, have postulated object-tracking visual indexes in the visual system, of which no computational account is given; the idea seems to be that they keep track of objects in a more primitive way, somewhat as one might keep track of a dog by keeping hold of its leash, without the keeping track being computationally based. But in the case of visually keeping track of a single object, it is possible to appeal to the spatial framework of vision so that we can appeal to the particularity of places in explaining how we keep track of the particularity of an object. We seem to have no analog of the spatial framework to appeal to in explaining how we keep track of someone's particular ongoing mental process. As we have just seen, even making fully explicit what the identity of an ongoing mental process consists in is not easy. But if we can do this, it will remain that we have no ready way of even approaching the question of how a social robot might go about following the particularity of an individual's ongoing mental process. That is, we have no ready way of explaining how to go about programming empathy—the ability to follow particular chains of thought and feeling—in an individual subject.

Our grasp of what's okay and what's not okay, what is all right to do or say and what isn't, is provided, in part, by our imaginative grasp of singular causal relations in one another's minds. It seems to be a further question how we should think of the assessments we make of one another's mental processes—or how a robot should go about making the kinds of assessment described by Collingwood. One picture would be that it is provided by our grasp of a general schema, or general schemas, for moral appraisal. But another picture would be that the moral assessment here is irredeemably particularist. That is, perhaps our empathic understanding of one another does generate assessments of one another, but those assessments are always specific to the particular contexts in which they are made and defy generalization.

5. RIGHTS FOR ROBOTS?

It has often been observed that just as people care about animals, so people care about machines. Or rather, there are animals that people care about, and there are some that we historically regard as merely commodities. Similarly, there are machines that people care about, and there are machines that we regard as merely means of production. The phenomenon with machines is very striking in the military. A marine sergeant running a robot repair shop in Iraq remembered one technician bringing a robot that was used to defused improvised explosive devices. "There wasn't a whole lot left of Scooby," Bogosh says. The biggest piece was its three-inch by four-inch head, containing its video camera. On the side had been painted "its battle list, its track record. This had been a really great robot." The veteran explosives technician looming over Bogosh was visibly upset. He insisted he did not want a new robot; he wanted Scooby-Doo back. "Sometimes they get a little emotional over it," Bogosh says. "Like having a pet dog. It attacks the IEDs, comes back, and attacks again. It becomes part of the team, gets a name. They get upset when anything happens to one of the team. They identify with the little robot quickly. They count on it a lot in a mission." (Garreau 2007). Soldiers will regard their robots with care and affection, sometimes giving them promotions. In a famous incident, the roboticist Mark Tilden developed a device like a five-foot-long stick insect for detonating land mines. When it prodded a landmine with one of its legs, that leg would get blown up. The device would then propel itself to the next landmine. During trials at Yuma, the thing worked beautifully, and it detonated five landmines before propelling itself with its one remaining leg to the last mine. At this point, the army colonel overseeing the trial "blew a fuse" and ordered the trial stopped. Tilden was devastated. But the colonel insisted it was "inhumane" and wouldn't let it go on. The phenomenon is not confined to trials. There are already well-documented accounts of the difficulty military personnel have in maltreating robots used for explosive ordnance disposal (Carpenter 2013, 2016), and it's not confined to military personnel. Ordinary humans, having been familiarized with social robots and then instructed to "torture" them, simply won't do it, and on being shown videos of people maltreating social robots, they exhibit many of the same negative reactions as they do to videos of people being maltreated (Darling, Nandy, and Brezeal 2015). Hitchbot was a robot with an enormous thumb that hitchhiked around America for a while, reporting its travels on Twitter and extending an enormous thumb to cars passing on the freeway. After many hundreds of hours of successful travel, Hitchbot was vandalized and utterly destroyed. Twitter exploded with fury at the people who could do such a thing (Wakefield 2019).

There is perhaps an element of play-acting in our response to social robots, in the same sense that there is an element of play-acting in our response to actors on a stage. One might argue that we ought to think in terms of a conception of distance in our reactions to BlabDroids, for example (Bullough 1912). In the theater, you do not want to become emotionally engaged with the action to the point that you leap onstage to tell Othello that he's being misled. But you also do not want to pull back from the action onstage to the point where you sit there thinking, "These people are all actors, and the person playing this character is pretending to give information to the person pretending to be that character." You calibrate your distance from the action onstage in such a way that you manage to resist leaping onstage yet are emotionally engaged with the characters and action. Just so, with social robots, you don't want to suppose that they literally are minded, yet you remain emotionally engaged with them enough to appreciate them. In these terms, perhaps the examples just given merely show that people are not very good at calibrating their distance from robots. The military robots drew the responses of care and concern by accident. That's not what they were designed for, and on the battlefield, it could be dangerous that they generate this reaction. Or it could be exploited. But social roboticists are ruthlessly and ingeniously designing machines to pull our reactions, probing reactions like trust, feelings of betrayal, the desire to help a needy robot, and so on. We have to be ready for an avalanche of robots that deliberately draw emotional responses because that's the point of the robot.

Consider sex robots. Consider sex robots that look like children. Is this okay? You might argue—and you might be right—that the ready availability of such robots would increase the number of attacks on human children. But it's also possible that the evidence could go the other way. Suppose it does. Suppose it turns out that the ready availability of child sex robots actually has no effect, or even a slight decrease, in the number of attacks on human children. Then is there anything problematic about such robots being available? One attitude is that "looking is free," as people in favor of the availability of pornography might argue. However, it's possible to argue that we as people would be damaged by this kind of thing. It's easy to imagine a society that says, "We don't do that stuff. That's not the kind of people we are." The argument here is not that this would have a bad effect on human children or that the robots themselves would suffer significant damage; we can presume that they are merely machines. The argument is rather that this is bad for the people themselves and for the societies they inhabit—that the practice would deform you as a person.

We seem to take a somewhat similar line in criminalizing the use of most recreational drugs. Kant had the idea that we have only "indirect" duties toward animals; as Calhoun puts it, "At the heart of [Kant's] indirect duty argument is a causal claim: Some actions regarding both persons and non-persons have the causal effect of weakening, destroying, or interfering with the cultivation of dispositions that aid us in the performance of our moral duties to persons" (Calhoun 2015). This kind of argument might apply equally well to machines as to animals.

If you are a fan of qualia, you may feel that this analogy doesn't work because the machines don't have qualia. And if you believe that the source of all value is in sensations of pleasure and pain, that may be cogent. We are only play-acting in our empathetic responses to machines if they don't have qualia.

But if the discussion to this point has been right, there is another consideration. Are the robots capable of singular psychological causation that is not grounded in general causation? If they are, they are appropriate targets of empathy by us, and they may be capable of empathizing with us. This is not the same as the question whether they have sensations. But a capacity for singular psychological causation not grounded in the general, and for empathy with the singular causation in human psychology, would seem to give robots a claim to membership in the human community that would be at least as strong as anything provided by merely having sensations.

Do the machines have responsibilities toward human beings? The classic answer to this, familiar to most high schoolers, is provided by Asimov's First Law of Robotics, "A robot may not injure, or, through inaction, allow a human being to come to harm." The trouble with this "law" is that it is virtually impossible to implement in a robot. For the thing to work, the robot would have to have some conception of what it is for a human to suffer injury or harm. It's often hard enough for humans to know when they are harming one another. You've suffered bereavement. Should I call on you, or do you really need time alone? I don't know, but I take a chance on it and call on you, to your great distress. If it's hard for me to know about this kind of thing, what chance does a social robot have? The concept of harm is not easy to program into a machine. Even the most stereotypical cases need heavy qualification. Removal of a healthy limb from a human being, for example, might cause harm—unless the human is out in the wilderness and trapped underneath a fallen rock, and removal of the limb is the human's only hope of escape. More generally, it's hard to see how there could be responsibilities that robots have toward humans, absent a kind of freedom and comprehension that robots do not at the moment seem likely to have.

CHAPTER FOUR

THE MIND-BODY PROBLEM

Let's consider how all the ideas we've been looking at bear on the mind-body problem. We'll begin with general causation and the interventionist approach we considered in Chapter 1. Then we'll look at how considerations of process and mechanism bear on relations between mind and body.

I. RELATIVITY TO A VARIABLE SET AND CAUSAL EXCLUSION

Recall the interventionist analysis of causation that we looked at in Chapter 1. In figure 4.1, we begin with the (possibly very large) set of variables characterizing the causal functioning of a system, such as the human mind, the human body, or an economy. We find a correlation between two variables, X and Y. We ask, "Is the explanation of this correlation that X causes Y?" With the interventionist approach, this is the same as the question "If there were to be an external intervention on X, would the values of X still be correlated with the values of Y?" For this approach to work, the intervention must meet certain conditions: the intervention must seize control of the variable X so that no confounding variable Z has any effect on X, the intervention must not be affecting the outcome directly (no placebo effects), and the intervention must not be correlated with any cause of the outcome (cf. Woodward 2003). The point I want to highlight here is that in this kind of approach, talk of causation is always relative to a variable set. This idea is actually ubiquitous in the causal modeling literature, and it fits with a broadly pluralistic approach to science and scientific explanation. There are some constraints on what constitutes a legitimate variable set:

1. To make the set of variables sufficiently rich. For example, we do not want to have an unrepresented common cause of any pair of variables in our variable set. Suppose there is an unrepresented common cause Z of two variables, X and Y, in our variable set. Suppose Z explains a correlation between X and Y. Then consider an intervention variable I that meets the conditions set out above, with respect to all the variables that are in the set. We could then find that X and Y are still correlated under the action of I. But that correlation is not explained by a causal connection between X and Y, but by the action of the hidden variable Z on both of them. Therefore, if we want the interventionist analysis to work at all, we have to ask that the variable set contain all common causes of any pair of variables in the set.

2. To keep the set of variables sufficiently lean. For example, the variables should be constitutively independent of one another. Each variable should stand for, in Hume's phrase, an "independent existent." Otherwise, we could find two

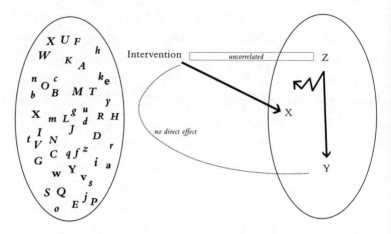

Figure 4.1. Variables X and Y are correlated. Whether X causes Y is determined by whether the values of X and Y are correlated under interventions on X. (For fuller explanation, see Chapter 1, §4.)

variables to be correlated under interventions even though the correlation was explained by their lack of independence rather than by a causal relation between them.

This pluralist perspective stands in contrast to a monistic, unificationist approach that takes Newton's *Principia* as its model. The idea here is that we have a relatively small set of basic principles on the basis of which everything else can be explained by being derived from the basic principles, perhaps also given further definitions and auxiliary facts about the world. With this way of thinking of it, the mind-body problem arises from the fact that there seems to be no way of deriving facts about the conscious life from facts about the physical world, and at this point the path branches.

1. We can claim to be able to find definitions of psychological terms in the vocabulary of basic physics, and thus we carry

through the Newtonian program even for the mind. All the operations of the mind can then be explained in terms of the functioning of a handful of basic physical principles, together with reductive definitions and auxiliary facts about the world.

Or:

2. We may find that it's impossible to carry through this reductionist exercise. In that case, the natural approach to the Newtonian program is to supplement the basic physical principles with principles relating to aspects of the mind—perhaps principles specifically concerning qualia, or whatever psychological characteristics seem basic to the rest. Then on this expanded basis, we can hope to provide a unified explanation of mental and physical phenomena.

Talking about the mind-body problem in the way that people usually do makes sense only if we have some monistic, unificationist conception of explanation, so that we have a way of saying what we are trying to solve when we try to solve the mind-body problem. The Newtonian picture seems to provide the kind of framework required. Here, for example, is Thomas Nagel with a classic statement of the mind-body problem:

The mind-body problem emerges in philosophy as the direct result of a modern ambition of scientific understanding—the desire to understand the world and everything in it as a unified system, so that the manifest diversity of natural phenomena is explained in terms of a much smaller number of fundamental principles. (Nagel 1993, 1)

Though there is certainly a role for unification in scientific explanation (e.g., Kitcher 1981), it is hard not to feel that the demand for unification is being pressed too hard here. Suppose we go back to the pluralism of the causal modeling approach, on which causation is always relative to a set of variables. We can deal with psychological variables, physical variables, or any combination of the two. But we can talk about causality only once we have relativized to a variable set. That means we give up on the idea of a single theory of everything that would explain, for example, both the behavior of gluons and the reasons for a rise in violence in a particular neighborhood. The demand for a single unified theory seems to have little force. It is arguable that even within physics, there is no attempt at unity across the board (Cartwright 1983). We can demand that within each set of variables to which we relativize causation, we make maximal and efficient use of the causal generalizations we find. However, there is no demand for a single set of variables relative to which all causal relations can be made explicit. Indeed, if we formulate interventionism by using an exogenous variable as our intervention, then it immediately follows that there can be no comprehensive specification of a list of variables as "all the causally significant variables." Such a totality of variables would, by definition, not include exogenous intervention variables. Therefore a totality of variables like that is not one relative to which we could find causal relations because of the impossibility of intervention on the variables in such a set (cf. Pearl 2000; Hitchcock 2007). This point is sometimes used as a vindication of Russell (1912): when we consider the entire universe, talk of causation simply drops away. This does not strike me as quite right, either about Russell or about explanation. Russell was arguing that the maturity of a science correlates with its abandonment of causality. The science in question might be highly domain specific, with no suggestion that it comprises all causally significant variables. Russell's idea seems to have been that once the science is at a point where we can find a comprehensive axiomatization of the variables that characterize the domain, then talk of causation drops away in favor of the axiomatization. But that doesn't require that the relevant set of variables be comprehensive, in the sense of including all variables that can be taken to have causal significance. The point that there is no such thing as a unique single set of variables to which all causation can be relativized remains.

Once we abandon the idea of a single set of fundamental principles from which all else can be derived and move to a pluralist picture of scientific explanation, the mind-body problem as it's usually conceived simply disappears. We no longer have to choose between saying that the fundamental principles are entirely physical (and that the mental principles are defined in terms of them) and that the fundamental principles include psychological laws, as the panpsychist says. Because there is no single set of fundamental principles.

Incidentally, there is another popular way of posing the mind-body problem: as a search for neural mechanisms of consciousness. This is a remarkably ill-posed question. There are neural mechanisms, as there are mechanisms in the physical world generally, and we have a good working, though rough, conception of what a physical mechanism is in terms of the transmission of motion by impulse or exchange of conserved quantities; neural mechanisms clearly fall under this general heading. I've been arguing so far that we also have a good working, though rough, conception of psychological processes and mechanisms: the idea of the sequences of psychological states that are meaningfully or rationally connected. We have not the dimmest idea of what we are talking about when we postulate a mechanism that is not of either of these types and that might somehow connect physical and mental. The quest for such a thing is not a hard scientific problem—it is not a problem at all, because it is entirely undefined.

The pluralist picture bears on Kim's well-known argument for epiphenomenalism, the idea that the mental never has causal significance. Kim diagrams the situation as shown in figure 4.2 (Kim 1998). We can't intervene independently on mental state A and physical state A; we can affect mental state A only by affecting physical state A. But then the situation seems open to interpretation.

Physical state A causes physical state B.
Mental state A supervenes on physical state A.
Mental state B supervenes on physical state B.
Mental state A does not cause mental state B; the mental states are
    epiphenomenal upon the underlying physical progression.

But mental state A and physical state A are not "independent existences," to use Hume's phrase. (If they were, then we could intervene on mental state A while leaving physical state A unchanged and determine whether mental state A causes mental state B.). Therefore the variable set here is not sufficiently lean. We cannot regard ourselves as dealing with a single variable set here.

There are different legitimate variable sets we could use: see, for example, figures 4.3 and 4.4. These are different variable sets. Relative to the first variable set, mental state A causes mental state B. Relative to the second variable set, physical state A causes

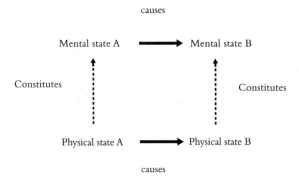

Figure 4.2. Kim diagram. (Jaegwon Kim, *Mind in a Physical World*, figure "Causes and Constitutes," © 1998 Massachusetts Institute of Technology, by permission of The MIT Press.)

mental state B. But we do not here have causal overdetermination. Causal overdetermination is when we have two sufficient causes for an effect within a single variable set, such as when we have both a bullet from the right striking the heart and a bullet from the left striking the heart. "Bullet from the right" and "bullet from the left" are independent variables relating to distinct existences. (See Woodward 2015 for a review of related analyses.)

2. OPTIMIZING VARIABLE SETS

There is a further question we can ask. Suppose that, as in the situation previously mentioned, we have two different variable sets, each of which can be used to characterize the causal functioning of a system. There is undeniably an impulse to say, "Yes, but which one—mental state A or physical state A—is the cause of mental state B?" One reaction to that is to say that it's simply missed the point of the relativization. We might compare Tarski's

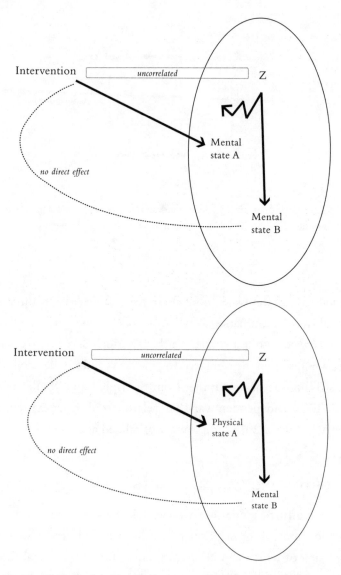

Figure 4.3 and *Figure 4.4.* Two different, possibly maximal variable sets that we can use to characterize causal relations for the same individual or individuals. Because Mental State A and Physical State A are not constitutively independent variables, we can't use them both in the same causal diagram. We get different answers to causal questions relative to the different causal diagrams.

idea that the truth predicate has to be relativized to a language. From Tarski's perspective, someone who insists that the sentence "This sentence is false" must be one or the other, true or false, is simply failing to recognize the language-relativity of the truth predicate, which means that the question is not well-posed. Similarly, someone who presses the question "But was the cause mental or physical, or both?" is failing to recognize the relativity of causation to a variable set, which means that the question is not well posed. But there is still a further question. Suppose we have two rival sets of variables in terms of which we can characterize our causal system. Suppose all the variables in each of the sets are legitimate. There still seems to be a question: But which variable set is better?

One way of responding to this question is to frame criteria for variable choice. The idea is that when those criteria are properly articulated, we will see that psychological variables turn out to be the best for characterizing the causal functioning of humans. There are a number of discussions of this topic in the literature. For example, Yablo (1995) argued that there should be a one-to-one map between the values of a cause variable and the values of an outcome variable. Suppose there are a number of different neural states that realize a psychological property such as "intending to buy a coffee." Yablo said, in effect: It doesn't matter which of these neural states realize the property. One way or another, if a person has the psychological property, whichever physical realization it has, then that will raise the probability of buying a coffee. If we characterize the cause variable at the level of neural state, we will find that there are many different values of the variable that lead to the same outcome. Therefore we should characterize the cause variable at the level of psychological state. Now, the demand for a one-to-one map does not sound quite right as it stands. Suppose you consider the relation between the air pressure in a tire and how well it grips the road. Perhaps as the air pressure goes up, the grip on the road improves, and then after a bit, as the air pressure further increases, the grip on the road goes down. It still seems evident that the right cause variable here is air pressure rather than some gerrymandered variable that would collapse together different values of air pressure so as to maintain a one-to-one map between cause and effect variables. Still, perhaps we can get something of the effect Yablo wanted by demanding that the map from cause to effect should be "locally one-to-one" in that neighboring values of the cause variable should be mapped to distinct values of the outcome variable.

There are many further criteria to be considered here. One approach is to look at Hill's (1965) famous "criteria for causation." Hill's criteria—for example, that we should look for a dose-response curve between cause and effect variables, that there should be big effects of variations in cause on the outcome variable, and so on—have a canonical place in the causation literature. But there is a puzzle about their status. Are they constitutive demands on what is required for a causal connection? Or are they empirically grounded in facts about some independently grounded conception of causation? They don't seem constitutive, but Hill does not provide any observational evidence for them. I think the right way to think of most of them is that they are criteria for the choice of the "best" variable set to characterize the causal functioning of a system. Consider the following function from x to y:

If x is rational, and z/y as its lowest expression, then $f(x) = y$.
Otherwise, $f(x) = 0$.

This function is concisely described, but we would require some convincing that it describes a cause-effect relationship between two variables. The problem is not that it isn't one-to-one, though it isn't. The problem is that for any arbitrarily brief interval of values of x, the values of y may be varying from arbitrarily low to arbitrarily high, with frequent resets to 0. It doesn't seem altogether impossible that there should be a causal relationship between two variables related like that. But this is not what Hill would have called a "dose-response" curve, where, for example, the probability of contracting cancer increases more or less smoothly with the quantity of cigarettes smoked. In looking for the best variable set to characterize the causal functioning of a system, we should be looking, so far as possible, for one where the cause variables have a dose-response relationship to the outcome variables.

Let's look at an example. Suppose it's announced that a prize will be given for finding the causes of schizophrenia. Suppose you have a scanner that gives a complete microphysical description of a human body, and suppose we scan thousands of subjects at various stages of their lives over a period of years. We observe which of those subjects develop schizophrenia and which do not. We will be able to form a big, disjunctive characterization of total microphysical states that are nomically sufficient for schizophrenia. This may not be an exhaustive disjunction—there may well be total microphysical states not on the list that would lead to the onset of the disease. But it may be true, for each of the microphysical states on the list, that some interventions on it would make a difference to whether schizophrenia is the outcome. Therefore we do have a variable set here relative to which we have identified causes of schizophrenia. But this was not what we were looking for when asked for identification of the causes of schizophrenia. One reason is that the approach is too fine-grained: we certainly do not have here a one-to-one map from microphysical states of the body to schizophrenia, because many differences between neural states will make absolutely no difference to the probability of contracting schizophrenia. Neither do we have any ready way of framing a dose-response relation between the cause variables and the outcome variables. But it would be possible to argue that these obstacles could be surmounted by framing complex variables defined in terms of our base class of microphysical descriptions. We could simply lump together all the total microphysical states that have as their outcome the same level of risk for schizophrenia, and then we could assign a number to each clump, one clump getting a higher number than another if the microphysical states in it lead to a higher probability of schizophrenia. But this exercise in gerrymandering has brought us no closer to finding the causes of schizophrenia. One further criterion for variable choice would be to demand that the cause variable should be manipulable by local processes. That is, it should be physically possible for some actual process to operate selectively on the putative cause variable, moving it systematically along its possible values. We need this condition to protect us against an artificial gerrymandering of the values of the putative cause variable so that lip service is paid to our other criteria, even though in practice there is no possibility of systematically varying the value of that variable.

Reviewing these criteria for the choice of a "best" variable set to characterize the causal functioning of a system suggests that they reflect our anthropocentric preferences; they reflect merely our practical interests in pursing questions about causation (Franklin-Hall 2016). That is doubtless correct, but it does not of itself mean that there is anything anthropocentric or pragmatic about causation itself. Relative to any one variable set, it is a still an entirely objective matter as to which variables are causes of which. It is only in finding the choice of a variable set that human interests come into play.

With all that said, there is undeniably something artificial about the whole enterprise here. We frame our abstract criteria for variable choice, and then we announce at the end that, to no one's surprise, psychological variables best meet the criteria. It is hard to shake the feeling that the thing has been somehow rigged, that we knew all along what the right answer would be. Then the question is: What explains our prior and deep commitment to the use of ordinary psychological variables in causal explanation?

I think at this point, we have to step outside the interventionist framework and consider how it goes for variable choice in the context of a process conception of causation. Consider the physical case. For example, Philip Dowe has argued in many places that we should understand physical causation in terms of "exchange of conserved quantities" (Dowe 1995, 323; Dowe 2000).

If this is the right way to characterize causal processes, then of course there is a special place for variables relating to conserved quantities in causal explanation. They describe physical causality at the most fundamental level. As Dowe acknowledges, there is indeed a question for this line of thought as to whether it can recognize the existence of high-level causation at all. Similarly, if our conception of a mental process is characterized in terms of the variables of ordinary, common-sense psychology, then of course there is a special place for those variables in mentalistic causal explanation. We saw the role of mentalistic thinking in connection with our imaginative understanding of singular causal connections. We saw, following Eells (1991), that there are certain independencies between singular causal facts and general causal facts. But still at the general level, we do need a conception of process or mechanism (cf. Chapter 1, Section 4 above). And because that is characterized in psychological terms, there is a certain primacy to those psychological terms in causal explanation. This makes sharp the question why we have the conception of mental process that we do. Dowe justifies his conception of physical process by saying that it's grounded in physical science. But the whole point of our approach has been to contrast science with the imaginative understanding characteristic of psychological explanation. Ultimately, we need to know what justifies our using the variables we do in imaginative understanding of one another. Notice also the peculiar status that this approach assigns to variables like intelligence (in the sense of IQ). These aren't variables that figure directly in the characterization of mental processes (unlike knowledge or intention). They seem more like theoretical constructs; they are certainly assumed to have causal significance. But they're not grounded in physical science; they're grounded in the variables used by imaginative understanding.

3. THE VALIDATION OF A PSYCHIATRIC CLASSIFICATION

There is a deeper question in play when we consider the mind-body problem. An insistence on unification in science can seem arbitrary, but there is a further question that, though related, does not depend on an uncritical unificationism. This is the issue of the validity of our ordinary psychological classifications. The idea of validity as it's usually used in connection with scientific constructs has two dimensions. One kind of case is when we have a phenomenon (some characteristic of objects such as mass or temperature) and want to know whether a particular way of identifying its presence or measuring it is any good. For example, we might all agree that there is such a thing as general intelligence in humans, and we have some idea of how it's caused and what differences it makes, but we argue about whether particular types of IQ tests provide good ways of quantifying it. Perhaps these tests might be challenged as subject to some cultural bias, so that results are affected by the specifics of one's general knowledge in a way that IQ itself is thought to be indifferent to. Here, the existence of the thing, intelligence, may not be in question, but particular ways of detecting or measuring it are up for assessment as more or less valid ways of measuring that thing. The other, more radical dimension of the idea of the validity of a construct has to do with whether there is anything there to detect or measure. For example, the category "neurasthenia" might be declared to be invalid not because any one measure of it is somehow incorrect but because there is no such thing at all. Similarly, in physical chemistry, the category "phlogiston" could be declared to be invalid not because any measure of it is incorrect but because it doesn't exist. This can happen even when we are not working with a fully explicit characterization of the variables in the relevant causal structure, but we are thinking of "phlogiston" as merely a latent variable playing a specified causal role. In this section, we'll look at how the question of validation and its relation to the mind-body problem play out in psychiatry; in the next section, we'll look at how it works for our ordinary psychological classifications. I think it's instructive to begin with psychiatry because paradoxically, it is sometimes somewhat easier to see the problems when one looks at the more complex and less familiar case.

Consider the *DSM-5* criteria for schizophrenia (American Psychiatric Association, 2013). There are five. To qualify for a diagnosis of schizophrenia, a patient must display at least two of the five symptoms, and one of them must be from the first three: (a) delusions, (b) hallucinations, (c) disorganized speech, (d) disorganized or catatonic behavior, and (e) negative symptoms (such as lowered cognitive functioning and flattened affect). Each of these symptoms is distressing, and one might reasonably think that each of them suggests that medical help is needed for the person displaying them. But what is the point of bundling them together in this way, as signs of a single condition? They seem quite heterogenous. Moreover, any two patients qualifying for a diagnosis of schizophrenia might do so on the basis of different subsets of the five criteria. So why should we think there is a single diagnostic category here?

The question is of a sort that is very familiar to philosophers: Is there such a thing as schizophrenia? Does schizophrenia exist? Perhaps the most popular answer is that it depends on whether schizophrenia is a "natural" characteristic (Lewis 1983), with the idea being that it is somehow up to science to determine what is natural (Lewis 1984). So how do the scientists determine whether there is such a thing as schizophrenia? Figure 4.5 shows the validators for the existence of a diagnostic category that were used in the preparation of *DSM-5* (Kendler et al. 2009, 27; for more recent discussion, see Appelbaum 2017).

Notice first that on the surface, this list proceeds entirely in terms of correlations. We are looking for what correlates we can find with someone having been established to have a disorder by means of a particular set of diagnostic criteria, such as the criteria for schizophrenia summarized earlier. The validity of the disor-

I Antecedent Validators
A. *Familial aggregation and/or co-aggregation (i.e., family, twin or adoption studies)
B. Socio-Demographic and Cultural Factors
C. Environmental Risk Factors
D. Proper Psychiatric History
II Concurrent Validators
A. Cognitive, emotional, temperament, and personality correlates (unrelated to the diagnostic criteria)
B. Biological Markers, e.g., molecular genetics, neural substrates
C. Patterns of Comorbidity
[Note - while categories A and B would most typically be assessed after illness onset, they also could be assessed prior to illness onset as pre-morbid characteristics]
III Predictive Validators
A. *Diagnostic Stability
B. *Course of Illness
C. *Response to Treatment

Figure 4.5. From the Guidelines for Making Changes to DSM-V used by the American Psychiatric Association, this is a summary list of the validators of a diagnostic category. High priority validators are marked with an asterisk. For more recent discussion, see Appelbaum (2017). (Kenneth S. Kendler, D. Kupfer, W. Narrow, K. Phillips. and J. Fawcett. 2009 "Guidelines for Making Changes to DSM-V." Washington, DC: American Psychiatric Association. Unpublished manuscript.)

der has to do with the number and strength of those correlations. And as remarked, there may be different weights given to different types of validators in this process. Most strikingly, there are quite different types of consideration being taken into account: the discovery of excessive synaptic pruning distinctively among schizophrenic patients would provide one type of validation of the category (see Sellgren et al. 2019 for discussion and further references), but so too are distinctive types of problems with social cognition (Penn et al. 1997).

Now, what is the point of this kind of exercise in validation? There are different ways in which we can think of the use of DSM criteria. One is roughly analogous to the way we think of criteria for giving someone a job, such as exam qualifications, commitment, relevant social skills, and so on. Here, we need not be thinking that there is one underlying condition to which all these indicators point. But it would be perfectly possible to assess the validity of a particular picture of a strong applicant by looking at the antecedent, concurrent, and predictive validators of the tests that are actually being applied by interviewers; indeed, something like that is how these tests actually are evaluated. On the other hand, you could be thinking of the disorder as something that is the causal outcome of the antecedent validators, which is expressed in the concurrent validators, and has as a causal outcome the way things go with the predictive validators. This is quite different to the case of criteria for a successful job application. On the face of it, it seems possible that one conception of the causal structure here may lead to a quite different weighting of validators than does the other conception of causal structure. It's hard to see how the discussion of validation in DSM can operate at the level of correlations without at least implicitly bringing in some picture of the causal structure of the disorder. Indeed, Ahn and Kim (2008) found that clinicians do operate with causal models of the disorders they are diagnosing, with factors that are thought to be responsible for many other aspects of the condition being given greater weight in diagnosis than those that are merely effects.

In fact, insofar as we have an interest in why disorders arise, and in understanding why some treatments work and some do not, there seems to be little choice but to regard the correlations used in validating disorders as pointing to causal networks that explain those correlations. We can understand those causal networks in terms of intervention counterfactuals, as Borsboom and Cramer (2013) suggest. But it's a natural thought that we may be able to go deeper and understand the essential cores of psychiatric disorders in terms of process and mechanism. The natural suggestion is that there will turn out to be some biological structure that is distinctive of schizophrenia, which is generated by the environmental and genetic risk factors, which is expressed in the concurrent symptoms, and which is addressed by the treatments that work. It's not impossible that there will turn out to be such a thing, though the complexity of the task is enormous (see Insel 2010 and Huckins et al. 2019 for optimistic assessments). But even given that we find such a biological structure, there are problems about understanding the relations between this structure and the psychological life of the patient in terms of process and mechanism, rather than merely intervention counterfactuals. In the physical case, we have a reasonably firm grasp of what constitutes a causal process: at bottom, it's something like what Locke (1690 / 1975) called "the transmission of motion by impulse," what Fair (1979) called the "transfer of energy," or what Dowe (2000) talks of in terms of "exchange of conserved quantities." Something like that is the basic idea of a physical process. Biological processes can generally be seen as particularly complex versions of underlying physical processes conceived in this way. In the psychological case, on the other hand, we think of mental processes as what we grasp by way of our imaginative understanding of the other person; it is our understanding of the other person "genetically by empathy," using Jaspers's phrase, that provides us with our grasp of a mental process—a psychological causal process. This is how we achieve our knowledge of how one psychic event emerges from another—of how Han's being attacked gives rise to his defensiveness, for example. The notion of mechanism stands to general causation somewhat as the notion of process stands to singular causation. If you're told that Sally's smoking caused her cancer, you can ask what the process was by which it did so. If you're told that in general, smoking causes cancer, you can ask, "What's the mechanism?" What you're looking for is something like the structure that sustains a particular type of process. So if we ask, "What's the mechanism by which smoking gives rise to cancer?" for example, we know how to go about addressing that. And if we ask, "What's the mechanism by which grief generates anger?" for example, we know how to go about giving an imaginative understanding of that. But when we ask, "What's the mechanism by which psychological and genetic factors might combine to produce a biological outcome?" or "What's the mechanism by which a biological structure might generate particular aspects of the conscious life?" we find that we have no idea as to how to answer. It's not that there are empirical facts we do not know. We simply do not have any relevant conception of process or mechanism in terms of which to frame the question. It's not that we have a sensible question here, with much work to be done to find the answer. It's rather that we do not know how to pose the question.

It is very often the case that we can establish general causation without knowing anything about the mechanism by which it works. Snow (1855) famously demonstrated that the water supply could be a cause of cholera while having only the haziest idea of the mechanism by which contaminated water produces cholera. For example, in a randomized controlled trial of a drug, to find whether it prevents breast cancer, if the trial is well executed, then it can provide knowledge of the causal connection even if the experimenters' conjecture as to the mechanism by which the drug is working turns out to be wrong.

This opens the possibility that, looking at trials across a population, we could find that both environmental and psychological factors and genetic factors were correlated with the outcome of a biological structure distinctive of schizophrenia. In fact, by exploiting the possibility of natural experiments, we could demonstrate a causal connection between these factors and schizophrenia as outcome. We could do this without having the slightest idea how to go about thinking of a mechanism by which these two factors might combine to generate the outcome.

Elisabeth, Princess of Bohemia, famously challenged Descartes as to whether there could be causal interactions between mind and body, given his dualist conception of mind and body as different substances (cf. Mattern 1978 for references and overview). As an undergraduate, this problem always struck me as fake; hadn't Hume (1740 / 1975) shown that causation was constant conjunction? There can be correlations between mental and bodily states, and there can be correlations under interventions between mental and bodily states. The problem arises when we think of causation in terms of mechanism and process. We know about mental mechanisms and processes, and we know about physical mechanisms and processes. But we have absolutely no understanding of how there could be mechanisms and processes linking mental and physical. My point is that the Princess Elisabeth problem is written large in psychiatry. It means that we do not have a clear conception of the causal structures our diagnostic procedures are trying to identify. If we think of causality in terms of mechanism and process, then we can't work with a mixed set of psychological and physical variables in specifying the causal structures we are trying to identify. But we have only optimism supporting the idea that we'll be able to give a purely psychological account of the relevant causal structures; similarly, it is only optimism that supports the idea that we'll be able to eliminate the mental and give a purely physical account of the relevant causal structures.

4. VALIDATION OF ORDINARY PSYCHOLOGICAL TERMS

How should we think of the validation of psychological terms? What does it take for there to be such things as love or fear? In the strongest view, we can't have a warranted belief about what's going on in someone's mind until we know the biological foundation of psychological states and that it validates our ordinary imaginative understanding (see fig. 4.6). In this kind of picture, a downward arrow indicates the direction of transmission of warrant. A sideways arrow pointing at a downward arrow is used to indicate that what's on the left of the arrow is a condition of the downward transmission of warrant (cf. Pryor 2012 for this notation). The present proposal might say, for example, that if you know that human pain is grounded in C-fiber firing, and you know that our imaginative understanding of one another provides, among other things, a good way of tracking C-fiber firing, then you have what it takes for your imaginative understanding of another person to warrant a belief that that person is in pain.

This approach immediately seems too strong to be a convincing analysis of what it takes for our ordinary imaginative understanding of one another to ground beliefs about other minds. At the moment, no one has any knowledge of an external biological perspective on the mind from which we could see how our ordi-

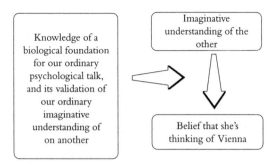

Figure 4.6. A view on which the validity of a psychological classification depends on our knowing a biological grounding for the classification.

nary talk of psychology is validated. Yet surely it's possible that we already do know about one another's mental states.

Perhaps a more convincing account would say not that we must know such a biological condition to be met but that such a biological condition must in fact be met, whether we know it or not, for our imaginative understanding of one another to generate warranted beliefs. So we might have what is shown in figure 4.7. In this picture, what is required is only that there be some such state as C-fiber firing, grounding our ordinary talk of pain, and that our imaginative understanding of one another does provide a way of tracking that biological condition. Again, for present purposes we need only a rough understanding of the notion of a biological ground for a psychological state.

This kind of picture is what ordinarily underlies the excitement and trepidation with which the mind-body problem is usually approached. This picture opens up the possibility that our ordinary talk of the mind could turn out to be ill-founded, in whole or in part. Investigations into neurobiology could, in principle,

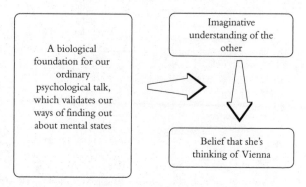

Figure 4.7. A view on which the validity of a psychological classification depends simply on there being a biological grounding for the classification.

show that there is no way of regarding our ordinary use of psychological categories as properly grounded in the biological, and therefore that by the present proposal, our imaginative understanding of one another doesn't warrant our beliefs about one another. Perhaps some quite radical recasting of our psychological categories, or abandonment of the psychological altogether, would be required. In that case, we would have lost all that we presently care about.

At this point, it will be evident that these are not the only ways we might think about the warranting of beliefs about other people's subjective lives. After all, the whole business of finding an external vantage point from which to assess our ways of finding out about some domain is often challenged. The proposals we have been considering take it that biology provides an external characterization of mentalistic phenomena, which we can use to assess our ways of finding out about the mind. But perhaps there is no such external vantage point to be had. An alternative diagram is shown in figure 4.8, with no substantive biological condition on the transmission of warrant. Quine famously said that there is no

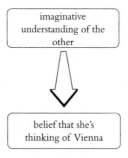

Figure 4.8. A view on which the validity of a psychological classification does not require any foundation in biology at all.

"first philosophy" firmer than science (Quine 1969). The point of this proposal is that equally, there is no first biology firmer than imaginative understanding. Given a way of knowing about the world, it's natural for anyone, not just philosophers, to look for a framework within which this way of knowing can be viewed in relation to the external environment and evaluated for its strengths and weaknesses. Quine's point about science is that there is no such external vantage point from which science can be assessed. Scientists can and do engage in radical critiques of current science, but that is not criticism from a standpoint external to science itself.

Similarly, we can and do criticize our empathetic understanding of other people. But where legitimate, that is not criticism from a standpoint external to the enterprise of imaginative understanding itself. It's a natural thought that perhaps we could cast natural science in the role of such an external standpoint from which our empathetic understanding of one another could be described and assessed.

The problem is that if we apply this approach to our empathetic understanding of other people, we will get the result that there is nothing in other humans that corresponds to the thoughts and feelings we reflexively ascribe to them. We have a scientific description of the human being we're trying to understand. We have a description of the external signs by means of which the human engages our imaginative understanding. And we have a cognitive-science description of imaginative understanding itself. None of this requires ascribing to the human we're trying to understand anything like thoughts and feelings.

In the case of hard science, as Quine pointed out, we don't generally admit the legitimacy of this kind of external criticism. There seems to be no reason why that point should be accepted for hard science but not for our imaginative understanding of one another. The general question that's raised here is whether our imaginative understanding of one another has any independent epistemic authority. Or does epistemic authority reside with science? Let's consider the sense in which science can be said to be epistemically authoritative about what's going on.

It's natural to oscillate between what we might call formal and concrete conceptions of science. On a formal, or empty, conception of science, anything at all that might constitute evidence for or against any proposition is counted as science. For a long time, I myself was skeptical about any argument about the limits of science, on the grounds that anything that could be established to exist, on whatever basis, would be grist for the mill. Scientists would be delighted if, for example, some nonphysical stuff could be established to exist—something not made out of anything like the currently recognized particles and forces—and they would devote considerable resources to finding out more about it. Science is, after all, merely the practice of thinking rationally about what's there. How could there be anything that lies outside its scope?

On the other hand, the fact is that we know fairly well what we're talking about when we talk about science. We're talking about the radical program of explaining nature in terms of measurable aspects of fundamental atoms and mechanical forces that was begun in the seventeenth century and continues today with the international system of scientific units, the fundamental forces and particles of the standard model, and the periodic table. The program has gone through a great deal of development since the seventeenth century, but it's recognizably the same program. On this concrete conception of science, there's absolutely no a priori reason to think that everything will be explainable in these terms, even if we expand or develop the program still further. At best, there is an induction: the success of this radical program has been vastly greater than could have been anticipated in the seventeenth century. Area after area has succumbed to the general approach. Therefore, all areas will succumb to the general approach. But of course, that induction is convincing only until we come upon a recalcitrant case. And consciousness is such a recalcitrant case. At the moment, we have no way of accessing phenomena of consciousness in terms of gluons and neurons. Our only access to the phenomena of consciousness is by way our imaginative understanding of one another.

We might draw a comparison with the question, "Why should I be moral?" On the one hand, in one formal conception of it, morality simply encompasses everything that might be reckoned as a reason for doing one thing rather than another, and the idea that there might be practical considerations outside the scope of morality makes no sense. On the other hand, we have a more concrete conception of morality in which we know perfectly well that it has a number of quite specific recommendations for action: that we should all treat each other well and fairly, that one shouldn't be mean, and so on. In this more concrete conception of it, morality is certainly open to critique—for example, as being no more than an expression of timidity and weakness. Given this concrete conception of it, it would certainly be possible to find directives for practical action that are outside of morality.

The concept of objectivity plays a significant role in people's thinking here. On the one hand, "objective" is simply a redundant qualifier of truth. Everything true is objectively true; that's what "objective" means. On the other hand, "objective" is sometimes used to characterize a method of inquiry and is contrasted with methods of inquiry that involve the use of imaginative understanding. Imaginative understanding involves taking up the other person's point of view. As such, it can't be said to be objective (for a seminal use of "objective" in this way, cf. Nagel 1974). In fact, sometimes the suggestion is that it's only science, in the concrete conception of it I just mentioned, that achieves objectivity. With this reading of objective, there's no reason to think that all truths are objective truths. The truths about someone's conscious life are still truths even if they can be accessed only by way of our imaginative understanding of them.

Even if we abandon the idea of finding an external foundation for common-sense psychology, or the idea that psychological terms must be validated by finding biological correlates for them, we can still pursue the mind-body problem. But everything looks a bit different. As it's usually stated, the mind-body problem is independent of questions about how we know about other people's mental states and our own. The way in which the mind-body problem is currently posed is usually in strictly ontological terms. The question is how the mind is related to the world of physics. Seen in these terms, the mind-body problem is simply an expression of a kind of ontological tidying up that tends to preoccupy philosophers but doesn't have much significance beyond that.

Consider the position of a devout theist who's also intrigued by the works of W. V. Quine. Quine's ontology comprises only physical objects and sets. "Is there room in Quine's ontology for God?" our theist wonders. God seems unlikely to be a physical object, but could God be a set? You might go quite far down this track of ontological tidying without ever having your commitment to theism remotely challenged. The epistemic credentials of your religious beliefs here are not being challenged. Nor are the epistemic credentials of your commitment to the physical objects and sets in Quine's ontology. You're simply wondering about the relation between them.

Another model is someone who wonders about the relation between the natural numbers and set theory. Are numbers sets? A person pursuing this question need never experience any threat to belief in the existence of numbers or belief in the existence of sets. You could view the mind-body problem in this light. We know about the existence of joy, sorrow, and so on. We also know about the existence of brain cells and so on. What's the relation, if any, between them? A good answer to this question might leave intact our understanding of both mind and body. The theist who successfully identifies God with a set might leave it at that, perhaps looking at theology in a somehow different light, but with little change to one's religious beliefs. Someone who reduces numbers to sets might not change any strictly arithmetical beliefs at all. Just so, you might manage to identify human passions with various brain states but leave your understanding of love and resentment otherwise unchanged.

5. THE BILLIARD BALLS

Suppose we have a relatively frictionless billiard table. The balls, once set in motion, will roll for months, perhaps years. We are going to set them all in motion with a cue applied to a single ball. And suppose that once set in motion, the balls on the table constitute a deterministic physical system. That is, from (a) the laws governing the system and (b) the initial positions of the balls and the vector describing the force, location, and direction of the initial cue shot, one can derive for each of the balls just where it will be at any subsequent time. If anything is a physical system, this is.

How many possible initial configurations of the balls are there? Lots; there is no particular restriction here on where we put them all. How many possible initial cue shots are there? Lots; we can apply the cue to any ball with any force or direction. Consider an arbitrary time after the initial cue shot—say, the following Tuesday at 2:00 p.m. How many possible configurations of the balls are there for that time? After doing the mathematics here, the answer is lots. Depending on the initial conditions and the initial cue shot, the balls may be in any random configuration the following Tuesday at 2:00 p.m.

Among all those possible configurations, there is one in which all the balls have congregated in the top left corner of the table on the following Tuesday at 2:00 p.m. There is no special significance to this; it is simply one among many random configurations that the balls may take for that time, depending on the initial conditions and the initial cue shot.

Consider now that the balls are rolling for months. Suppose that for a given initial condition and cue shot, we look over a period of fifteen weeks at where all the balls are each Tuesday at 2:00 p.m. In general, we won't find any pattern here. But suppose we find a particular initial condition and cue shot for which, each Tuesday at 2:00 p.m., all the balls congregate in the top left corner of the table. Again, there is no special significance to this. It is simply one among many random configurations we might find at that time, depending on the initial conditions and the cue shot. It is the special case of a random configuration in which there is repetition of the configuration and the configuration is briefly describable.

Suppose I do the initial setup of the balls and the cue shot in just this way, and to my astonishment, over the next fifteen weeks, I observe this pattern. "Why do they all collect in the top left corner on Tuesdays at 2:00 p.m.?" I ask. "What is the cause?" Now there is a sense in which there is a cause of the balls all collecting like this, and I know it already—it was the initial setup together with my cue shot—but that does not answer my question. I am wondering whether there is some cause of the balls all collecting in the top left corner rather than the bottom left corner; and whether there is some cause of their all collecting on Tuesdays at 2:00 p.m. rather than on Mondays at 2:00 p.m. There may be no cause variable whose values are systematically correlated with the place and time of congregation of the balls. It is for this reason that we say it is merely random that the balls happen to congregate at that time and place. If, for example, I systematically manipulate the force of the cue shot, I do not make systematic differences to the place and time of congregation. Most changes to the force of the cue shot lead to there being no congregating of the balls at all. We do not have any systematic relation between the basic physical variables in terms of which the billiard balls and cue shot are characterized on the one hand, and the place and time of congregation on the other hand.

Any philosopher reading this will reflexively suggest the possibility of gerrymandering a complex physical variable, a big disjunction of more specific variables characterizing the initial conditions and cue shot, which one could regard as systematically related to the place and time of congregation. For the moment, my only comment on this proposal is that we would not regard the construction of such a gerrymandered variable as discovering the cause of the pattern. Consider a real case: the documented tendency of eels to congregate in the Sargasso Sea. What's the cause? Why congregate there rather than anywhere else? The kind of factor that is usually suggested is some use of geomagnetic patterns in navigation by all the eels. An ingenious philosopher might proceed by another route: you might give a more basic physical description of the eels and their physical environment and claim that this specifies the cause of the eels meeting at one place rather than another. In fact, although we might applaud the ingenuity, we would regard it as wasted: you have not discovered the cause of the eels meeting in one place rather than another. And suppose you did manage to face us down so that no one could think of a good reason to complain about the gerrymandered variable. You would produce only dismay because you would have destroyed what seems like a perfectly good and important distinction: the distinction between the congregation of the billiard balls, which is an accident, and the congregation of the eels, which is not an accident.

This explains the sense in which even in a deterministic physical system, there can be a physical outcome without a physical cause. There can be patterns that appear by accident; there is no causal explanation of why we have this pattern rather than another related pattern. There is no causal explanation for why the billiard balls are congregating at that particular place and time rather than some other place and time. It is in that sense that their congregating at that place and time is merely random, only an accident, or an epiphenomenon. The randomness, in this sense, is entirely consistent with this being a deterministic physical system. The point is emphatically not that there are nonphysical forces at work.

To sum up the point so far, when we ask, "Why are the billiard balls congregating at (p, t)?" we are implicitly asking, "Why are they congregating at (p, t) rather than at $(p1, t1)$, $(p2, t2)$, and so on?" The sense in which there is no cause for their congregation at that place and time rather than any other is that there is no variable that could be manipulated to systematically vary the place and time of congregation. Contrast the case if all the balls had magnetic cores, and we had a moveable electromagnet under the table. Then we could say, "The balls are congregating at the top left rather than any other place (or are not congregating at all) because the electromagnet is switched on at under the table at the top left." Had the magnet been switched on at the bottom right, the balls would have been at the bottom right, and so on. But we don't have any such variable in the case of the billiard balls. That's the sense in which there isn't a cause.

Suppose now that we consider not billiard balls but people. Let us suppose, though, that people are physical systems. In fact, let us suppose that they are deterministic physical systems, just as our billiard balls are, and although they have psychological states, those psychological states supervene on the underlying physical reality. Once you have fixed the physical facts, you have fixed all the psychological facts.

Suppose we have a group of people, such as the members of a seminar or a committee. They agree to meet weekly for the fifteen weeks of term at a particular place, the Dennes Room, and a particular time, Tuesday at 2:00 p.m. Every week, they congregate at that place and time.

Suppose you don't know that these organisms are sentient, but you do have a complete physical description of them. Suppose that you are a "Martian physicist," in Nozick's sense: you know all about the physics of human beings, can take it in at a glance, and can compute physical implications of the current situation in a moment. The physics of human beings is no more complex for you than the physics of a billiard table is for humans. At the most fundamental physical level, you know all about the physical environment, and you know all the relevant physical laws. Because we are dealing with a deterministic physical system, you will be able to predict the movements of all these things, so you will be able to predict that they will congregate at that place and time. But why do they congregate at that place and time rather than any other? Well, you say, it's random. It's just an accident. There is no particular reason why they are congregating at that place and time rather than any other. There is no physical variable whose values are systematically correlated with variations in the place and time of meeting. An ingenious philosopher might point out that in principle, you could come up with a gerrymandered physical variable whose values are related to variation in place and time of meeting, but you would not allow that as a demonstration that this is not an acci-dental pattern. You would not allow that kind of ingenuity to de-stroy the contrast between this case and the case of the congrega-tion of the eels at one place rather than another, because as with the eels, this is really not an accident.

What about the situation of an ordinary person observing the class who does know something about the psychologies of the peo-ple involved? This observer knows that everyone in the group agreed to meet in the Dennes Room on Tuesdays at 2:00 p.m. That is why they all keep coming. This observer therefore knows a variable that is systematically related to the time of congregation: having agreed to meet at place p and at time t. Systematic variation in the value of this variable is correlated with changes in the place and time of meeting. There are, for example, natural disasters that will disrupt the seminar, and the correlation would break down if, for instance, everyone had in the seminar had foolishly agreed to meet on a mountaintop at midnight. Nonetheless, we would ordinarily say that we know why we are meeting where and when we do.

In this situation, we have a deterministic physical system com-prised of all those organisms and their environment. Although the organisms have psychological states, those states supervene on the basic physics. We have a physical outcome: everyone is congre-gating at a particular place and time. There is a psychological cause for this congregation, the initial agreement, but it has no physical cause. From a physical point of view, the congregation is simply an accident.

I think that we do feel a strong resistance to the idea that the explanatory story might end there. We tend to feel strongly that in the case I described, we should look for some brain variable that is the cause of everyone meeting as they do. Somewhere in the pattern of electrical activity across the brain, there must be something recognizable as a diary. There must be some physical variable—perhaps a complex physical variable relating to the activity of massive assemblies of cells—whose systematic manipulation would make systematic changes to where and when everyone met. (By saying it may be complex, I don't mean that it will be gerrymandered in the sense I discussed earlier, but that it may be far removed from being simply the firing pattern of a single cell.) There must be the electric diary in the brain. Much research on the brain is built around the idea that it should be possible to find such complex physical variables. I don't want to resist the idea that we should search for these complex brain variables, but I do want to reflect on the status of the idea that these physical variables must exist, and on what difference it makes to our thinking about psychology whether they exist.

In our example, why should we think that there must be a (non-gerrymandered) physical variable whose values are systematically related to the time at which everyone meets? Notice that the existence of such a variable is not a consequence of the fact that we are dealing with a deterministic physical system in which all the psychological facts supervene on physical facts. We have already seen that there seems to be no contradiction in the idea that such a deterministic system might have psychological causation without physical causation. Or is the existence of such a physical variable somehow already implicated in the very idea that we are dealing with a causal relation? If we already have a psychological cause for a phenomenon, why should we think that there must be a physical cause? This is the right way to put the idea with which we began: that "the space of reasons" might be different to "the space of causes." It's not that things that stand in normative relations to one another should not be thought of as causal. It's rather that the space of causal psychological explanation may be orthogonal to the space of causal physical explanation.

REFERENCES

Ahn, Woo-Kyoung, and Nancy S. Kim. 2008. "Causal Theories of Mental Disorder Concepts." *Psychological Science Agenda* 22:3–8.

Allen, Keith. 2016. *A Naïve Realist Theory of Color.* Oxford: Oxford University Press.

American Psychiatric Association. 2013. *Diagnostic and Statistical Manual of Mental Disorders.* 5th ed. Washington, DC: American Psychiatric Association.

Appelbaum, Paul S. 2017. "DSM-5.1: Perspectives on Continuous Improvement in Diagnostic Frameworks." In *Philosophical Issues in Psychiatry IV: Psychiatric Nosology,* edited by Kenneth R. Kendler and Josef Parnas, 392–402. Oxford: Oxford University Press.

Austin, John L. 1957. "A Plea for Excuses." *Proceedings of the Aristotelian Society* 57:1–30.

Benessia, Alice, Silvio Funtowicz, Mario Giampietro, Angela Guimaraes Pereira, Jerome Ravetz, Andrea Saltelli, Roger Strand, and Jeroen P. van der Sluijs. 2016. *The Rightful Place of Science: Science on the Verge.* Tempe, AZ: Consortium for Science, Policy and Outcomes.

Bishop, John. 1981. "Peacocke on Intentional Action." *Analysis* 41:92–98.

Blaisdell, Aaron P., K. Sawa, K. J. Leising, and M. R. Waldmann. 2006. "Causal Reasoning in Rats." *Science* 311:1020–22.

Blaisdell, Aaron P., and M. R. Waldmann. 2012. "Rational Rats: Causal Inference and Representation." In *Handbook of Comparative*

Cognition, edited by E. A. Wasserman and T. R. Zentall, 175–198. Oxford: Oxford University Press.

Boghossian, Paul A., and J. David Velleman. 1989. "Color as a Secondary Quality." *Mind* 98:81–103.

Borsboom, Denny, and Angélique O. J. Cramer. 2013. "Network Analysis: An Integrative Approach to the Structure of Psychopathology." *Annual Review of Clinical Psychology* 9:91–121.

Bortolotti, Lisa. 2016. "Epistemic Benefits of Elaborated and Systematized Delusions in Schizophrenia." *British Journal for the Philosophy of Science* 67:879–900.

Browning, S. M., and S. Jones. 1988. "Ichthyosis and Delusions of Lizard Invasion." *Acta Psychiatrica Scandanavica* 78:766–67.

Bullough, E. 1912. "'Psychical Distance' as a Factor in Art and as an Aesthetic Principle." *British Journal of Psychology* 5:87–98.

Byrne, Alex. 2006. "Color and the Mind-Body Problem." *Dialectica* 60: 223–44.

Calhoun, Cheshire. 2015. "But What about the Animals?" In *Reason, Value and Respect: Kantian Themes from the Philosophy of Thomas E. Hill, Jr.*, edited by Mark Timmons and Robert M. Johnson, 194–212. New York: Oxford University Press.

Campbell, John. 2007. "An Interventionist Approach to Causation in Psychology." In *Causal Learning: Psychology, Philosophy, and Computation*, edited by Alison Gopnik and Laura J. Schulz, 58–66. Oxford: Oxford University Press.

———. 2020. "Does That Which Makes the Sensation of Blue a Mental Fact Elude Us?" In *The Routledge Handbook of Philosophy of Colour*, edited by Derek Brown and Fiona MacPherson. London: Routledge / Taylor and Francis.

Carpenter, Julie. 2013. "Just Doesn't Look Right: Exploring the Impact of Humanoid Robot Integration into Explosive Ordnance Disposal Teams." In *Handbook of Research on Technoself: Identity in a Technological Society*, edited by Rocci Luppicini, 609–36. Hershey, PA: Information Science Reference.

———. 2016. *Culture and Human-Robot Interaction in Militarized Spaces: A War Story.* Abingdon and New York: Routledge / Taylor and Francis.

Cartwright, Nancy. 1983. *How the Laws of Physics Lie.* Oxford: Oxford University Press.

Chang, Hasok. 2004. *Inventing Temperature: Measurement and Scientific Progress.* New York: Oxford University Press.

Chekhov, Anton. 1886. "Misery." *Peterburgskaya Gazeta* 26, January 16.

Cohen, L. Jonathan. 1981. "Can Human Irrationality Be Experimentally Demonstrated?" *Behavioral and Brain Sciences* 3:317–31.

Collingwood, R. G. 1959. "History as Re-enactment of the Past Experience." In *Theories of History: Readings from Classical and Contemporary Sources*, edited by Patrick Gardiner, 251–62. Glencoe, IL: Free Press.

Conan Doyle, Arthur. 1917. "The Adventure of the Cardboard Box." In *His Last Bow: A Reminiscence of Sherlock Holmes*. London: George H. Doran.

Costello, E. Jane, Scott N. Compton, Gordon Keeler, and Adrian Angold. 2003. "Relationships between Poverty and Psychopathology: A Natural Experiment." *Journal of the American Medical Association* 290: 2023–29.

Darling, Kate, Palash Nandy, and Cynthia Brezeal. 2015. "Empathic Concern and the Effect of Stories in Human-Robot Interaction." *Proceedings of the 24th IEEE International Workshop on Robot and Human Communication (ROMAN)*, 2015, 770–75.

Darwall, Stephen. 1974. "Pleasure as Ultimate Good in Sidgwick's Ethics." *Monist* 58:475–89.

Davidson, Donald. 1980. "Actions, Reasons and Causes." In *Essays on Actions and Events*. Oxford: Oxford University Press.

Dennett, Daniel. 1981. "True Believers: The Intentional Strategy and Why It Works." In *Scientific Explanation*, edited by A. F. Heath, 150–67. Oxford: Oxford University Press.

Dowe, Philip. 1995. "Causality and Conserved Quantities: A Reply to Salmon." *Philosophy of Science* 62:321–33.

———. 2000. *Physical Causation*. Cambridge: Cambridge University Press.

Eells, Ellery. 1991. *Probabilistic Causality*. Cambridge: Cambridge University Press.

Eliot, George. 1861. *Silas Marner*. London: William Blackwood and Sons.

Fair, David. 1979. "Causation and the Flow of Energy," *Erkenntnis* 14:219–50.

Fodor, Jerry. 1998. *Concepts: Where Cognitive Science Went Wrong*. New York: Oxford University Press.

Foot, Philippa. 2001. *Natural Goodness*. Oxford: Oxford University Press.

Frankfurt, Harry G. 1971. "Freedom of the Will and the Concept of a Person." *Journal of Philosophy* 68:5–20.

Franklin-Hall, Laura R. 2016. "High-Level Explanation and the Interventionist's 'Variables Problem.'" *British Journal for the Philosophy of Science* 67:553–77.

Freud, Sigmund. 1909. "Notes upon a Case of Obsessional Neurosis." *SE* 10:151–318.

Garreau, Joel. 2007. "Bots on the Ground." *Washington Post,* May 6.

Glymour, Clark. 2004. "We Believe in Freedom of the Will so That We Can Learn." *Behavioral and Brain Sciences* 27:661–62.

———. 2007. "When Is a Brain Like the Planet?" *Philosophy of Science* 74:330–47.

Glymour, Clark, and David Danks. 2007. "Reasons as Causes in Bayesian Epistemology." *Journal of Philosophy* 104:464–74.

Good, I. J. 1961–1962. "A Causal Calculus I–II." *British Journal for the Philosophy of Science* 11:305–318; 12:43–51.

Gopnik, Alison, Clark Glymour, David M. Sobel, Laura E. Schulz, Tamar Kushnir, and David Danks. 2004. "A Theory of Causal Learning in Children: Causal Maps and Bayes Nets." *Psychological Review* 111:3–32.

Gopnik, Alison, and Andrew Meltzoff. 1997. *Words, Thoughts and Theories.* Cambridge, MA: MIT Press.

Gopnik, Alison, and Henry M. Wellman. 2012. "Reconstructing Constructivism: Causal Models, Bayesian Learning Mechanisms, and the Theory Theory." *Psychological Bulletin,* 138, 1085–108.

Gordon, Robert. 1986. "Folk Psychology as Simulation." *Mind and Language* 1:158–71.

———. 1995. "Simulation without Introspection or Inference from Me to You." In *Mental Simulation: Evaluations and Applications—Reading in Mind and Language,* edited by Martin Davies and Tony Stone, 53–67. Oxford: Blackwell.

Griffiths, Thomas L., Falk Lieder, and Noah D. Goodman. 2015. "Rational Use of Cognitive Resources: Levels of Analysis between the Computational and the Algorithmic." *Topics in Cognitive Science* 7:217–29.

Grunbaum, Adolf. 1990. "'Meaning' Connections and Causal Connections in the Human Sciences: The Poverty of Hermeneutic Philosophy." *Journal of the American Psychoanalytic Association* 38:559–77.

Hall, Ned. 2004. "Two Concepts of Causation." In *Causation and Counterfactuals,* edited by John Collins, Ned Hall, and L. A. Paul, 181–204. Cambridge, MA: MIT Press.

Hansen, Katherine, Margaret Gerbasi, Alexander Todorov, Elliott Kruse, and Emily Pronin. 2014. "People Claim Objectivity After Knowingly Using Biased Strategies." *Personality and Social Psychology Bulletin* 40:691–99.

Hart, Herbert L. A., and Tony Honoré. 1985. *Causation in the Law.* 2nd ed. Oxford: Oxford University Press.

Hegel, Georg W. F. 1820 / 1991. "Preface." In *Elements of the Philosophy of Right,* edited by Allen W. Wood, translated by H. B. Nisbet, 9–23. Cambridge: Cambridge University Press.

Heider, Fritz, and Marianne Simmel. 1944. "An Experimental Study of Apparent Behavior." *American Journal of Psychology* 57:243–59.

Hill, Austin Bradford. 1965. "The Environment and Disease: Association or Causation?" *Proceedings of the Royal Society of Medicine* 58:295–300.

Hitchcock, Christopher. 1995. "The Mishap at Reichenbach Fall: Singular vs. General Causation." *Philosophical Studies* 78:257–91.

———. 2001. "A Tale of Two Effects." *Philosophical Review* 110:361–96.

———. 2007. "What Russell Got Right." In *Causation, Physics and the Constitution of Reality,* edited by Huw Price and Richard Corry. Oxford: Oxford University Press.

Honderich, Ted. 1982. "The Argument for Anomalous Monism." *Analysis* 42:59–64.

Huckins, Laura M., Amanda Dobbyn, Douglas M. Ruderfer, Gabriel Hoffman, Weiqing Wang, Antonio F. Pardiñas, Veera M. Rajagopal, Thomas D. Als, Hoang T. Nguyen, Kiran Girdhar, James Boocock, Panos Roussos, Menachem Fromer, Robin Kramer, Enrico Domenici, Eric R. Gamazon, Shaun Purcell, CommonMind Consortium, The Schizophrenia Working Group of the Psychiatric Genomics Consortium, iPSYCH-GEMS Schizophrenia Working Group, Ditte Demontis, Anders D. Børglum, James T. R. Walters, Michael C. O'Donovan, Patrick Sullivan, Michael J. Owen, Bernie Devlin, Solveig K. Sieberts, Nancy J. Cox, Hae Kyung Im, Pamela Sklar, and Eli A. Stahl. 2019. "Gene Expression Imputation across Multiple Brain Regions Provides Insights into Schizophrenia Risk." *Nature Genetics* 51:659–74.

Hume, David. 1748 / 1975. "Enquiry concerning Human Understanding." In *Enquiries concerning Human Understanding and concerning the Principles of Morals,* edited by L. A. Selby-Bigge, 3rd ed. revised by P. H. Nidditch. Oxford: Oxford University Press.

Hursthouse, Rosalind. 1991. "Arational Actions." *Journal of Philosophy* 88:57–68.

Insel, Thomas R. 2010. "Rethinking Schizophrenia." *Nature* 486:187–93.

Jackson, Frank, and Philip Pettit. 1990. "Program Explanation: A General Perspective." *Analysis* 50:107–17.

Jaspers, Karl. 1913 / 1959. *General Psychopathology.* Manchester: Manchester University Press.

Jeong, Kwangmin, Jihyun Sung, Haesung Lee, Aram Kim, Hyemi Kim, Chanmi Park, Youin Jeong, JeeHang Lee, and Jinwoo Kim. 2018. "Fribo: A Social Networking Robot for Increasing Social Connectedness through Sharing Daily Home Activities from Living Noise Data." In *HRI 18: Proceedings of the 2018 ACM / IEEE International Conference on Human-Robot Interaction,* 114–22.

Kant, Immanuel. 1997. *Lectures on Ethics.* Edited by Peter Heath and Jerome B. Schneewind, translated by Peter Heath. Cambridge: Cambridge University Press.

Kendler, Kenneth S., John M. Hettema, Frank Butera, Charles O. Gardner, and Carol A. Prescott. 2003. "Life Event Dimensions of Loss, Humiliation, Entrapment, and Danger in the Prediction of Onsets of Major Depression and Generalized Anxiety." *Archives of General Psychiatry* 60:789–96.

Kendler, Kenneth S., and L. Karkowski-Shuman. 1997. "Stressful Life Events and Genetic Liability to Major Depression: Genetic Control of Exposure to the Environment?" *Psychological Medicine* 27:539–47.

Kendler, Kenneth S., L. M. Karkowski, and Carol A. Prescott. 1999. "Causal Relationship between Stressful Life Events and the Onset of Major Depression." *American Journal of Psychiatry* 156:837–41.

Kendler Kenneth S., D. Kupfer, W. Narrow, K. Phillips, and J. Fawcett. 2009. "Guidelines for Making Changes to DSM-V." Unpublished manuscript. Washington, DC: American Psychiatric Association.

Kendler, Kenneth R., Henrik Ohlsson, Dace S. Svikis, Kristina Sundquist, and Jan Sundquist. 2017. "The Protective Effect of Pregnancy on Risk for Drug Abuse: A Population, Co-Relative, Co-Spouse, and Within-Individual Analysis." *American Journal of Psychiatry* 174: 954–62.

Kim, Jaegwon. 1998. *Mind in a Physical World: An Essay on the Mind-Body Problem and Mental Causation.* Cambridge, MA: MIT Press.

Kitcher, Philip. 1981. "Explanatory Unification." *Philosophy of Science* 48:507–31.

Knutti, Reto. 2008. "Why Are Climate Models Reproducing the Observed Global Surface Warming So Well?" *Geophysical Research Letters* 35:L18704.

Kushnir, Tamar, Fei Xu, and Henry M. Wellman. 2010. "Young Children Use Statistical Sampling to Infer the Preferences of Other People." *Psychological Science* 21:1134–40.

LeDoux, Joseph E., and Daniel S. Pine. 2016. "Using Neuroscience to Help Understand Fear and Anxiety: A Two-System Framework." *American Journal of Psychiatry* 173:1083–93.

Lee, Geoffrey. 2019. "Alien Subjectivity and the Importance of Consciousness." In *Blockheads!: Essays on Ned Block's Philosophy of Mind and Consciousness*, edited by Adam Pautz and Daniel Stoljar, 215–42. Cambridge, MA: MIT Press.

Leshner Alan I. 1997. "Addiction Is a Brain Disease, and It Matters." *Science* 278:45–47.

Lewis, David. 1983. "New Work for a Theory of Universals." *Australasian Journal of Philosophy* 61:343–77.

———. 1984. "Putnam's Paradox." *Australasian Journal of Philosophy* 62: 221–36.

Locke, John. 1690 / 1975. *An Essay Concerning Human Understanding*. Edited by P. H. Nidditch. Oxford: Oxford University Press.

Martin, C. B., and Max Deutscher. 1966. "Remembering." *Philosophical Review* 75:161–96.

Maugham, W. Somerset. 1933. *Sheppey. A Play in Three Acts*. London: Heinemann

McDowell, John. 1994. *Mind and World*. Cambridge, MA: Harvard University Press.

Menzies, Peter. 2017. "Counterfactual Theories of Causation." In *Stanford Encyclopedia of Philosophy* (Winter 2017 Edition), edited by Edward N. Zalta. https://plato.stanford.edu/archives/win2017/entries/causation-counterfactual.

Michotte, Albert. 1963. *The Perception of Causality*. Translated by T. R. Miles and E. Miles. New York: Basic Books.

Moore, G. E. 1903. "The Refutation of Idealism." *Mind* 48:433–53.

Mori, M. 2005. "The Uncanny Valley." Translated by K. F. MacDormand and T. Minato. *Energy* 7:33–35.

Morin, Charles M., R. R. Bootzin, D. J. Buysse, J. D. Edinger, C. A. Espie, and K. L. Lichstein. 2006. "Psychological and Behavioral Treatment of Insomnia: Update of the Recent Evidence (1998–2004)." *Sleep* 29:1398–414.

Nagel, Thomas. 1971. "Brain Bisection and the Unity of Consciousness." *Synthese* 22 (1971): 396–413.

———. 1974. "What Is It Like to Be a Bat?" *Philosophical Review* 83: 435–50.

———. 1986. *The View from Nowhere*. Oxford: Oxford University Press.

———. 1993. "What Is the Mind–Body Problem?" In *Experimental and Theoretical Studies of Consciousness*, edited by G. R. Bock and James L. Marsh, 1–7. Ciba Foundation Symposium 174. Chichester: John Wiley and Sons.

Newton, Isaac. 1687. *Philosophiæ Naturalis Principia Mathematica*. London: Jussu Societatis Regiæ ac Typis Josephi Streater.

Nida-Rümelin, Martine. 1996. "Pseudonormal Vision: An Actual Case of Qualia Inversion?" *Philosophical Studies* 82:145–57.

Pan, Matthew K. X. J., Elizabeth A. Croft, and Günter Niemeyer. 2018. "Evaluating Social Perception of Human-to-Robot Handovers Using the Robot Social Attributes Scale (RoSAS)." In *HRI 18: Proceedings of the 2018 ACM / IEEE International Conference on Human-Robot Interaction,* 443–51.

Parfit, Derek. 1984. *Reasons and Persons.* Oxford: Oxford University Press.

Peacocke, Christopher. 1979. *Holistic Explanation: Action, Space, Interpretation.* Oxford: Oxford University Press.

Pearl, Judea. 2000. "Epilog: The Art and Science of Cause and Effect." In *Causality: Models, Reasoning and Inference.* Cambridge: Cambridge University Press.

Penn, D. L., P. W. Corrigan, R. P. Bentall, J. M. Racenstein, and L. Newman. 1997. "Social Cognition in Schizophrenia." *Psychological Bulletin* 121:114–32.

Phillips, Elizabeth, Xuan Zhao, Daniel Ullman, and Bertram F. Malle. 2018. "What Is Human-like?: Decomposing Robots' Human-like Appearance Using the Anthropomorphic roBOT (ABOT) Database." In *HRI 18: Proceedings of the 2018 ACM / IEEE International Conference on Human-Robot Interaction,* 105–113.

Pryor, James. 2012. "When Warrant Transmits." In *Wittgenstein, Epistemology and Mind: Themes from the Philosophy of Crispin Wright*, edited by Annalisa Coliva, 269–303. Oxford: Oxford University Press.

Quine, W. V. O. 1960. *Word and Object.* Cambridge, MA: MIT Press.

———. 1969. "Epistemology Naturalized." In *Ontological Relativity and Other Essays.* New York: Columbia University Press, 69–90.

Repacholi, Betty M., and Alison Gopnik. 1997. "Early Reasoning about Desires: Evidence from 14- and 18-Month-Olds." *Developmental Psychology* 33:12–21.

Rorty, Richard. 1979. *Philosophy and the Mirror of Nature.* Princeton: Princeton University Press.

Russell, Bertrand. 1912. "On the Notion of Cause." *Proceedings of the Aristotelian Society* 7:1–26.

———. 1948. "Analogy." In *Human Knowledge: Its Scope and Limits*. London: George Allen and Unwin, 482–86.

———. 1967. "Prologue: What I Have Lived For." In *The Autobiography of Bertrand Russell*. London: George Allen and Unwin, 9–10.

Rutter, Michael. 2007. "Proceeding from Observed Correlation to Causal Inference: The Use of Natural Experiments." *Perspectives on Psychological Science* 2:377–95.

Schaffer, Jonathan. 2016. "The Metaphysics of Causation." In *Stanford Encyclopedia of Philosophy* (Fall 2016 Edition), edited by Edward N. Zalta. https://plato.stanford.edu/archives/fall2016/entries/causation-metaphysics.

Sellars, Wilfrid. 1956. "Empiricism and the Philosophy of Mind." In *Minnesota Studies in the Philosophy of Science*, edited by H. Feigl and Michael Scriven, 253–329. Minneapolis: University of Minnesota Press.

Sellgren, Carl M., Jessica Gracias, Bradley Watmuff, Jonathan D. Biag, Jessica M. Thanos, Paul B. Whittredge, Ting Fu, Kathleen Worringer, Hannah E. Brown, Jennifer Wang, Ajamete Kaykas, Rakesh Karmacharya, Carleton P. Goold, Steven D. Sheridan, and Roy H. Perlis. 2019. "Increased Synapse Elimination by Microglia in Schizophrenia Patient-Derived Models of Synaptic Pruning." *Nature Neuroscience* 22:374–85.

Shah, Nishi, and J. David Velleman. 2005. "Doxastic Deliberation." *Philosophical Review* 114:497–534.

Shakespeare, William. 1603 / 2016. *Othello*. Edited by Ayanna Thompson and E. A. J. Honigmann. New York: Bloomsbury.

Shoemaker, Sydney. 1984. "Identity, Properties and Causality." In *Identity, Cause and Mind*. Cambridge: Cambridge University Press.

Sinai, Yakov G. 1970. "Dynamical Systems with Elastic Reflections: Ergodic Properties of Dispersing Billiards." *Russian Mathematical Surveys* 25:137–89.

Snow, John. 1855. *On the Mode of Communication of Cholera*. London: John Churchill.

Sober, Elliot. 1985. "Two Concepts of Cause." *Proceedings of the Biennial Meeting of the Philosophy of Science Association* 2:405–24.

Spirtes, P., C. Glymour, and R. Scheines. 1993. *Causation, Prediction and Search*. New York: Springer-Verlag.

Stark, Philip B. 2018. "No Reproducibility Without Preproducibility." *Nature* 557:613.

Stark, Philip B., and Andrea Saltelli. 2018. "Cargo-Cult Statistics and Scientific Crisis." *Significance* 15:40–43.

Surtees, P. G., P.McC. Miller, J. G. Ingham, N. B. Kreitman, D. Rennie, and S. P. Sashidharan. 1986. "Life Events and the Onset of Affective Disorder: A Longitudinal General Population Study." *Journal of Affective Disorders* 10:37–50.

Taylor, Alex H., Rachael Miller, and Russell D. Gray. 2012. "New Caledonian Crows Reason about Hidden Causal Agents." *Proceedings of the National Academy of Sciences of the United States of America* 109: 16389–91.

Thagard, Paul, and Richard E. Nisbett. 1983. "Rationality and Charity." *Philosophy of Science* 50: 250–67.

Tversky, Amos, and Daniel Kahneman. 1981. "The Framing of Decisions and the Psychology of Choice." *Science* 211:453–58.

Wakefield, Jane. 2019. "Can You Murder a Robot?" *BBC News,* https://www.bbc.com/news/technology-47090174.

Williams, Bernard. 1970. "Deciding to Believe." In *Problems of the Self.* Cambridge: Cambridge University Press, 136–51.

Woodward, James. 2003. *Making Things Happen: A Theory of Causal Explanation.* Oxford: Oxford University Press.

———. 2015. "Interventionism and Causal Exclusion." *Philosophy and Phenomenological Research* 91:303–47.

Woodward, James, and Christopher Hitchcock. 2003. "Explanatory Generalizations, Part 1: Counterfactual Account." *Nous* 37:1–24.

Xu, Fei, and Vashti Garcia. 2008. "Intuitive Statistics by 8-Month-Old Infants." *Proceedings of the National Academy of Sciences of the United States of America* 105:5012–15.

Xu, Fei, and Tamar Kushnir, eds. 2012. *Advances in Child Development and Behavior 43: Rational Constructivism in Cognitive Development.* Waltham, MA: Academic Press.

Yablo, Stephen. "Mental Causation." 1992. *Philosophical Review* 101: 245–80.

Youyou, Wu, Michal Kosinski, and David Stillwell. 2015. "Computer-Based Personality Judgments Are More Accurate Than Those Made By Humans." *Proceedings of the National Academy of Sciences of the United States of America* 112:1036–40.

ACKNOWLEDGMENTS

This book began as the Whitehead Lectures at Harvard in 2009. At the time, I wished I had made more progress with the topic; I now think I have, but it took a while. I learned a lot from Ned Hall's comments, both at the time and subsequently. My interest in the subject had begun somewhat earlier. In 2003–2004, I had spent the academic year at the Center for Advanced Study in the Behavioral Sciences in Stanford. Through the center, I worked with Alison Gopnik, who was arguing that causal modeling approaches could be used to analyze causal learning in children, and the statistician Thomas Richardson, who talked us through recent work on the relation between statistics and causality. I also talked a great deal with Jim Woodward, from whom I learned an enormous amount, both from his writings and in person over many years. Jim subsequently provided amazingly helpful and detailed comments on the penultimate draft of this book. I learned from discussions with Clark Glymour and Chris Hitchcock. Through the center, I met the psychiatrist Ken Kendler, who has had a big impact on my thinking, providing a very rich collection of examples of psychological causation and ways of thinking about the explanation of psychiatric phenomena. In the fifteen years or so since then, we've talked by phone fortnightly, and I've learned an enormous amount from him.

In developing these ideas, I have benefited enormously from having had the chance to present this material in a series of graduate seminars in Berkeley. I particularly remember exchanges with Austin Andrews, Adam Bradley, Jennifer Coluccio, Peter Epstein, Jim Hutchinson, Alex Kerr, John Schwenkler, Umrao Sethi, James Stazicker, and Klaus Strelau.

Imogen Dickie gave me extremely detailed and helpful feedback on a (much longer) early draft. Bill Brewer and Susanna Siegel provided detailed comments on the penultimate draft that were incredibly helpful. I've had a lot of discussions over many years with colleagues: Barry Stroud, Quassim Cassam, Geoff Lee, Naomi Eilan, Christoph Hoerl, Teresa McCormack, Johannes Roessler, Wes Holiday, Seth Yalcin, and Alva Noe. Ned Block provided interesting comments at an APA panel on causation. Mike Martin engaged vigorously with this material, particularly through his sharp comments on a long series of presentations in a seminar in fall 2018. The mathematician Ivan Smith gave me a lot of help in thinking through criteria for variable choice in causal explanation. The child psychiatrist George Downing provided me with new perspectives on the explanation of psychological phenomena. I've also been lucky to talk with artist and roboticist Alex Reben, who introduced me to the topic of social robots and continues to provide challenging ideas; we've figured as cosymposiasts in a variety of contexts, and I've always learned from him.

Since 2012, I've cotaught a class called Sense and Sensibility and Science with a social psychologist, Rob MacCoun, and a physicist, Saul Perlmutter; the psychologist Tania Lombrozo joined more recently. The topic of causation came up extensively, and they really shaped my thinking on the subject.

Also, I am grateful to audiences at meetings or colloquia in Copenhagen; Open AI in San Francisco; Harvard; Indiana; Columbus, Ohio; and Simon Fraser University. It was particularly helpful to be invited to give a master class at Fribourg, where Gianfranco Soldati and his colleagues gave the opportunity for an extended development of these ideas.

Finally, Cassandra Chen showed me how to do the art and otherwise inspired. Thanks Cassandra!

INDEX